Studies in Logic
Volume 14

New Approaches to Classes and Concepts

Volume 5
Incompleteness in the Land of Sets
Melvin Fitting

Volume 6
How to Sell a Contradiction. The Metaphysics of Inconsistency
Francesco Berto

Volume 7
Fallacies – Selected Papers 1972-1982
John Woods and Douglas Walton, with a Foreword by Dale Jacquette

Volume 8
A New Approach to Quantum Logic
Kurt Engesser, Dov M. Gabbay and Daniel Lehmann

Volume 9
Handbook of Paraconsistency
Jean-Yves Béziau, Walter Carnielli and Dov Gabbay, editors

Volume 10
Automated Reasoning in Higher-Order Logic. Set Comprehension and Extensionality in Church's Type Theory
Chad E. Brown

Volume 11
Foundations of the Formal Sciences V: Infinite Games
Stefan Bold, Benedikt Löwe, Thoralf Räsch and Johan van Benthem, editors

Volume 12
Second-Order Quantifier Elimination: Foundations, Computational Aspects and Applications
Dov M. Gabbay, Renate A. Schmidt and Andrzej Szałas

Volume 13
Knowledge in Flux. Modeling the Dynamics of Epistemic States
Peter Gärdenfors. With a foreword by David Makinson

Volume 14
New Approaches to Classes and Concepts
Klaus Robering, editor

Studies in Logic Series Editor
Dov Gabbay   dov.gabbay@kcl.ac.uk

New Approaches to Classes and Concepts

Edited by
Klaus Robering

© Individual author and College Publications 2008. All rights reserved.

ISBN 978-1-904987-85-7

College Publications
Scientific Director: Dov Gabbay
Managing Director: Jane Spurr
Department of Computer Science
King's College London, Strand, London WC2R 2LS, UK

http://www.collegepublications.co.uk

Original cover design by Richard Fraser
Created by orchid creative   www.orchidcreative.co.uk
Printed by Lightning Source, Milton Keynes, UK

All rights reserved. No part of this publication may be reproduced, stored in a retrieval system or transmitted in any form, or by any means, electronic, mechanical, photocopying, recording or otherwise without prior permission, in writing, from the publisher.

Preface

The present volume contains the contributions to the workshop "Alternative Set Theories — Alternatives to Set Theory" held at the University of Southern Denmark at June 2nd, 2006. The workshop was intended to be a forum for the discussion of philosophical, foundational, and computational aspects of non-standard set-theories (as, for instance, Ackermann's class theory with unrestricted comprehension, Gilmore's "Nominalistic Logic", and Aczel's Theory of "unfounded sets") — and, of course, alternatives to set-theory (as, for instance, Klaus Grue's "Map Theory"). It brought together researchers from computer and information science, mathematics, philosophy, and history of science to discuss the concept of a *set* and that of a *class*, the role of these concepts in the foundation of mathematics, their relationships to the traditional distinction between extension and intension, and the ontological and practical issues connected with them and their application in theoretical computer science and other areas of application (as, for instance, linguistic pragmatics). A brief account of the contributions and their interrelationships is given in the introduction to this volume.

I would like to thank Anne Gerdes, former director of IT-Vest[1], and Lotte Weilgaard, head of the Institute of Business Communication and Information Science at the University of Southern Denmark, for supporting the workshop by their grants. Lotte Weilgaard, furthermore, was so kind to host the workshop at her institute.

Kolding, August 2007 *Klaus Robering*

---

[1] IT-Vest is a consortium of the Universities of Southern Denmark, Aalborg, and Aarhus as well as of the Aarhus School of Business to promote and support higher education in the fields of computer science and information technology.

Contents

Introduction
Klaus Robering 1

Fixing Cantor's Paradise: The Prehistory of Ernst Zermelo's Axiomatization of Set Theory
Volker Peckhaus 11

Ackermann's Class Theory
Klaus Robering 23

Nominalistic Logic: From Naive Set Theory to Intensional Type Theory
Jørgen Villadsen 57

A Gentle Introduction to Map Theory
Klaus Grue 87

ε-Style (of) Semantics
Sebastian Bab, Bernd Mahr, Tina Wieczorek 111

Assertions in AFA Set Theory
Anton Benz 143

Index 177

List of Contributors

Sebastian Bab
Berlin Institute of Technology
Faculty IV (Electrical Engineering
and Computer Sciences)
FR 6–10
Franklinstr. 28/29
Berlin, D-10587
bab@cs.tu-berlin.de

Anton Benz
Centre for General Linguistics (ZAS)
Schützenstr. 18
Berlin, D-10117
benz@zas.gwz-berlin.de

Klaus Grue
Department of Computer Science
University of Copenhagen
Universitetsparken 1
Copenhagen Ø, Dk-2100
grue@diku.dk

Bernd Mahr
Berlin Institute of Technology
Faculty IV (Electrical Engineering
and Computer Sciences)
FR 6–10
Franklinstr. 28/29
Berlin, D-10587
mahr@cs.tu-berlin.de

Volker Peckhaus
Institute for Humanities: Philosophy
University of Paderborn
Warburger Str. 100
Paderborn, D-33098
volker.peckhaus@upb.de

Klaus Robering
Institute of Business Communication
and Information Science
University of Southern Denmark
Engstien 1
Kolding, Dk-6000
robering@sitkom.sdu.dk

Jørgen Villadsen
Department of Informatics and
Mathematical Modeling
Technical University of Denmark
Richard Petersens Plads
Building 321
Kongens Lyngby, Dk-2800
jv@imm.dtu.dk

Tina Wieczorek
Berlin Institute of Technology
Faculty IV (Electrical Engineering
and Computer Sciences)
FR 6–10
Franklinstr. 28/29
Berlin, D-10587
wieczo@cs.tu-berlin.de

Introduction

Klaus Robering

Institute of Business Communication and Information Studies
University of Southern Denmark

robering@sitkom.sdu.dk

The present volume is a collection of articles dealing with the notions of *set* and *class*, thus with the basic notions of set theory. However, emphasis is not so much on the technical development of set theory as a branch of mathematics but rather on foundational issues. This includes, of course, the role these notions play in the foundations of mathematics. Besides this, however, the contributions to the present volume take a more comprehensive view and discuss more broadly the ontology and epistemology of sets and classes, their role in concept formation, and their application in such fields as (theoretical) computer science and linguistics.

Classifying objects — thus collecting them into classes — is an operation both fundamental and common in everyday life as well as in scientific endeavours. Logic, philosophy, mathematics, and computer science analyse this activity by means of such notions as 'class', 'set', and 'type'. The systematic study of these concept has been inaugurated in the 19th century by philosophers, logicians, and mathematicians like Bolzano, Boole, Schröder, Frege, Dedekind, Cantor and others. Their efforts were continued in the early 20th century by researchers like Russell, Zermelo, Fraenkel, Skolem and von Neumann. Ultimately, their work culminated in the axiomatic formulation of classical set theory in different versions such as ZFC or NBG. The development leading from Cantor's informal set theory to Zermelo's celebrated axiomatization is described and analysed in Volker Peckhaus's contribution to this volume (p. 11–19). Being in one way a step towards orthodoxy, Zermelo's axiomatization involved both ideas transcending standard attitudes (cp. the last section of Peckhaus's contribution) and ideas used as starting points towards alternatives to set theory (cp. the introductory section of Klaus Grue's contribution, p. 86–110).

Zermelo's application of the axiomatic method was of course also a reaction to the so-called "paradoxes of set theory". And undoubtedly, this problem of the paradoxes has been one of the main sources of inspiration for the development of alternative set theories and alternatives to set theory. The question, however, what this "problem of the paradoxes" really consists in turns out to be intricate. Note that Peckhaus in his contribution speaks of "unintended" rather than of "paradoxical" sets, and John Myhill is cited in Robering's contribution as having attributed to Gödel the view that there never have been any such things as paradoxes of set theory (see p. 26). Of course, whether the "paradoxes" are justly called thus depends on the underlying conception of a class (or set) which one adopts.

As might be expected for a workshop like that documented in the present collection, quite different views of sets and classes have been at issue in its discussions and are therefore reflected in one way or other in the contributions to the present volume. One may recognize at least three different conceptions in the present articles: sets and classes

(A) are modelling tools,
(B) provide a foundation[1] (for mathematics and computer science),
(C) are basic concepts of logic.

The first view is alluded to in the subtitle of the contribution by Sebastian Bab, Bernd Mahr, and Tina Wieczorek (p. 111–140); and the approach developed in their paper is an alternative to modelling with orthodox sets. Many structures met with in specific application areas exhibit such features as self-referentiality, self-applicability, and circularity which are hard or even impossible to cope with by means of traditional sets. Bab, Mahr, and Wieczorek cite the functions of the λ-calculus, intensional meanings, and truth predicates as examples. Benz in his contribution (p. 143–176) adds yet another example, namely the pragmatics of common knowledge and common learning effects in dialogic communication. Benz as well as Bab/Mahr/Wieczorek are concerned with structures which are at odds with the axiom of foundation and apply non-standard tools — Aczel's AFA set theory and ϵ-sets, respectively — for coping with these structures.

Set theory as a foundation is at issue in the contributions of Peckhaus and Grue. Peckhaus describes the role of axiomatized set theory for both Hilbert's foundational program and Zermelo's own position in the early 1930s, which is opposed "to almost all foundational positions at that time"; see p. 19 below. Klaus Grue gives an introduction — and at the same time a new axiomatization — of his map theory (MT) which competes with Zermelo's set theory as a foundation of mathematics. Indeed, map theory is even more ambitious: "[...] where ZFC is suited as a foundation of mathematics, MT is suited as a foundation of both mathematics and computer scienc"; see p. 87 below.

Grue's map theory is not an alternative set theory, but it is an alternative to set theory. Traditional set theory starts from sets and reduces functions *via*

---

[1] ZF (or ZFC) has become something like the "standard foundation" for mathematics. A précis of the historical development leading to this situation is provided by Villadsen in his contribution to the current volume, cf. especially p. 66 f. Villadsen gives an account of both the most important technical results in connection with this developments and the parallel developments in information technology.

the concept of a pair to sets.² This procedure has been especially criticized by category theorists who argue that the relevant properties of mathematical objects derive from their abstract structure rather than from their constitution out of parts. Whereas the latter is described by set theory by means of its membership relation, the former has to be analyzed in terms of functional mapppings relating different objects to each other. Thus a foundation of mathematics based on the notion of mapping would be much more "natural" than one which starts from "collections". Let this as it might be³, here it suffices to note that Grue's map theory rests on the side of functions. Furthermore, these functions are computable — as it seems natural for a theory aiming at a foundation of computer science.

Perhaps, one may say that "sets as foundations" (the second item of our list above) are just the modelling tools of the mathematician; furthermore, that what satisfies the mathematician's need may not be expected also to suit the needs of other scientists — *e.g.*, the linguist (cp. Benz's contribution) or the computer scientist (cp. the contributions of Grue and Bab/Mahr/Wieczorek). However, what is adequate for the mathematician's need? As already hinted at above, this is well a matter of debate — or rather one of several debates involving different parts. One should be aware of the fact that the issue of "sets as foundations" involves more than the question for adequate modelling tools for the mathematician. What is at stake here is first of all the matter of legitimization. One of the earliest criticisms of set theory has been the constructivistic one put forward from different, though related angles of views, as for instance that of the French "semi-intuitionists" (*e.g.*, Borel (1905), Poincaré (1908, 2nd book)), that of predicativism (Weyl, 1918, 1–17), and that of Brouwer's full-blown intuitionism (Brouwer, 1918). The question at issue here is not so much that of the adequateness of certain (set-theoretic) notions but that of their legitimization. Even if it were possible to demonstrate by some kind of consistency proof that "modelling by classical sets" does not run into contraditictons, this would for Brouwer not demonstrate the legitimacy of, for example, the higher cardinalities. As contentual considerations irrefutably demonstrate to him: "Von anderen mathematischen Mächigkeiten, als die abzählbare, die abzählbar unfertige, und die continuierliche, kann gar keine Rede sein [There can be no question of other mathematical cardinalities

---

² Admittedly, this is only the (now) accepted view. Frege's theory of classes treats classes as extensions of concepts (see p. 6f) which in turn are special functions. Thus it "reduces" classes to functions. Furthermore, von Neumann's set theory was originally a theory of functions: "Wir ziehen aber vor, nicht die 'Menge', sondern die 'Funktion' zu axiomatisieren. Der letztere Begriff umfaßt ja gewiß den ersteren. (We prefer, however, to axiomatize not 'set' but 'function'. The latter notion certainly includes the former)"; von Neumann (1925, 396).

³ A discussion of this issue may be found in the first part of Feferman (1977).

besides the countable, the countable-unfinished and the continous]" (Brouwer, 1909, 571).[4]

Leaving alone the difficult question of justifying a theory about sets, let us return to the (perhaps) more modest question what kind of theory would be adequate for the need of mathematicians and other scientists. In their survey article on mathematical logic, Hans Hermes and Heinich Scholz call the first-order theory comprising just the unrestricted axiom of comprehension and the axiom of extensionality the *ideal calculus* (German: "Idealkalkül"; cf. Hermes and Scholz 1952, 57).[5] This label suggests that such a theory would fulfil its function and serve the needs of science in an ideal way. Of course, the *ideal calculus* is inconsistent, but in its simplicity it put forwards two fundamental ideas about sets — namely, that sets are (1) completely determined by their elements and that (2) each condition statable in the theory's language determines a corresponding set of elements satisfying this condition. The *ideal calculus* does not exclude the kind of circular and self-referential structures used by Benz and Bab/Mahr/Wieczorek to model speakers and self-applicable functions. However by Russell's (and Zermelo's, see p. 11) construction, it is easily seen to be inconsistent.

The above requirement (1) — thus the axiom of extensionality — provides for the criterion of identity for traditional sets. Fraenkel (1927, 67) in his famous introductory lectures on set-theory calles it "the axiom of determinateness". He explains that, given the usual properties of the identity relation, this axiom implies that predicates do not distinguish between sets having the same elements (since they are identical):

> Nach dem Axiom besitzt die hier eingeführte Gleichheit im vollen Maße die Eigenschaft, die ihr auch sonst überall in der Mathematik zukommt, daß nämlich je zwei als gleich erklärte Objekte — hier: Mengen — unterschiedslos einander vertreten können (According to this axiom, the thus introduced identity relation does completely possess that property, which it also has everywhere else in mathematics, namely that two obejcts — here: sets — declared to be equal may represent each other without making any difference).
>
> Fraenkel (1927, 67f)

---

[4] In another article, Brouwer stresses the importance of the issue of legitimacy by drawing a moral parallel to it. A consistency proof, he explains, does not yield anything of mathematical value: "[...] eine durch keinen widerlegenden Widerspruch zu hemmende unrichtige Theorie ist darum nicht weniger unrichtig, so wie eine durch kein reprimierendes Gericht zu hemmende verbrecherische Politik darum nicht weniger verbrecherisch ist (an incorrect theory, even if it cannot be inhibited by any contradiction that would refute it, is none the less incorrect, just as a criminal policy is none the less criminal even if it cannot be inhibited by any court that would curb it)"; Brouwer (1923, 3).

[5] Thus, the ideal calculus is a formalization of naive set theory; compare on this Villadsen's contribution, esp. p. 60.

Probably one would be more cautious today on extensionality than Fraenkel is in this quotation from 1927. Extended parts of recursion theory, for instance, are intensional by dealing with representations (codes) of functions rather than with the extensional objects encoded. Other examples of intensionality[6] may be found, for example, in the collection *Intensional Mathematics* (Shapiro, 1985), which by the way contains two article (by J. Myhill and N. D. Goodman) on "intensional set theory". Finally, let me quote Feferman (1975, 78) stating: "The mathematical role of extensionality [...] is much less important than ordinarily thought".

Grue's map theory is like ZFC "except that it begins with computable functions" (p. 87 below). Thus its follows the *ideal calculus* on the issue of extensionality. The treatment of extensionality is only "slightly more complicated" (p. 96) than in ZFC proper since we have to adapt extensionality to the functional point of view. The basic idea, however, is quite obvious: operations returning the same values for the same arguments should yield the same value if they themselves are taken as arguments by the same operation. On the other hand, the theories of Ackermann and Gilmore presented in the contributions by Villadsen (p. 57–85) and Robering (p. 23–54) are lacking extensionality; and this seems to be quite typical for set theories trying to retain the unrestricted comprehension principle of the *ideal calculus*; cp. Feferman (1977, 1982).

AFA set theory, which is applied by Benz, may be conceived of as giving more internal structure to sets than ZFC. Like the latter, it retains extensionality from the *ideal calculus*. The question, however, arises how the identity criterion provided by extensionality is to be applied in view of the additional internal structure of sets. This is nicely described by Benz in his exposition of AFA set theory. Orthodox set theory identifies sets with the same members. Given the axiom of foundation this leaves only poorly structured entities behind. For his special purpose, however, Benz needs a "conception of a set as a structured object where the identity of objects follows from the identity of its constitutents, *i.e.*, elments" (p. 148 below). As an appproach to modelling, the ϵ-style semantics[7] of Bab/Mahr/Wieczorek eschews decisions on the existence and identity of sets (which go beyond that what is guaranteed by classical predicate logic plus identity). However, the issue of extensionality returns in the form of distinctions. A set is distinguished from its extension, *i.e.*, the collection of its members. Orthodox extensional set theory identifies these two items, but one is by no means forced to do this and may admit "intensional sets".[8]

---

[6] On the role of intensionality in mathematics and natural language cp. furthermore Villadsen's contribution; esp. pages 73 and 80.

[7] Thus ϵ-style semantics should be seen in the larger context of a general theory of modelling; cp. Mahr (2003) concerning such a theory.

[8] The whole issue of extensionality within ϵ-style semantics is discussed in Section 4 of the contribution of Bab/Mahr/Wieczorek.

The notion of an extension is of course well known from logic and has a tradtion reaching as far back as to Aristotle's doctrine of genus and species developed in his *Categories*. Commenting on Porphyry's introduction (*Isagoge*) to Aristotle's text, Cajetan (*i.e.*, Thomas de Vio, 1469–1534) explains that each general term may be either conceived of intensionally (*modo intensive*) or extensionally (*modo extensive*); Cajetan (1939, chap. 1, part 6). The intensional view is concerned with the general term's conceptual content whereas the extensional view regards the individuals to which the term is truly applied. This view has been codified in the doctrine that general terms have both an intension and an extension, the intension being the concept expressed by the predicate, the extension the class of objects to which the concept applies.

Sets then are just the extensions of concepts and set theory becomes a part, or perhaps rather an extension, of the logical theories of concepts. Note that this concept of a set differs markedly both from the cumulative-iterative concept implicit in Zermelo's view[9] and the notion of a set as an "internally structured object". Aczel (1980, 31) calls it the "naive" or "logical" notion of a set; I will call the thus conceived sets "classes" since it is more common in philosophy to speak of the extension of a concept as the class rather than the set of all objects to which the concept applies.[10] This conception is to be found in the works of Frege, of course, but it also makes its appearance, for instance, in the writings of Weyl, cf. Weyl (1918, 1966, §4, I.2, resp.).[11] Frege uses classes for the sake of his logicistic project; thus we have reached here item (C) — sets as logical concepts — of our list above. These lines of thought are reflected in this collection by the contributions of Villadsen (on Paul Gilmore's neologicistic set theory) and Robering (on Ackermann's typefree logic). Given the interpretation of classes as extensions of concepts, the comprehension axiom of the *ideal calculus* just states that every concept has an extension. The axiom of extensionality furthermore fixes the characteristic trait of extensions —

---

[9] Cp. the well-known exposition of that conception by Boolos (1971).

[10] Of course, that use of the term *class* should not be confused with the technical use of this term in set theories distinguishing (proper) classes from sets.

[11] Usually, Frege speaks of the "course of values" rather than the extension of a concept. Occasionally, however, he uses "Umfang", the German equivalent to "extension". In his review of Schröder's VORLESUNGEN ÜBER DIE ALGEBRA DER LOGIK, he explains that a class is the extension of a concept, but does by no means consist "aus den Gegenständen, die unter den Begriff fallen [...], sondern er hat an dem Begriff selbst und nur an diesem seinen Halt (of the objects to which the concept applies, rather it is only supported by that very concept and by nothing else)"; Frege (1895, p. 455). In a similar vein, Weyl (1966, 26) opposes against a collective conception of sets: "Die Vorstellung, daß eine solche Menge durch Kolligieren aus ihren einzelnen Elementen zusammengebracht wäre, ist durchaus fernzuhalten. Daß wir die Menge kennen, soll nichts anderes besagen, als daß uns eine für ihre Elemente charakteristische Eigenschaft gegeben ist (The idea that such a set is put together by collecting its individual elements is to be avoided. That we know the set says nothing else than that we know a property characteristic for its elements)."

namely: to be extensional. Villadsen and Robering deal with different, though in some respects similar ways to uphold unrestricted comprehension, thus the principle that each concepts has an extension. However, both systems lack extensionality.[12]

Within the logical conception of a class, the question arises of course what kind of object the extension of a concept is. Platonism has been one option here; and quite a few interpreter of Frege have argued that he has taken this route though the exegetical issue is not quite clear.[13] However, there is a whole plethora of alternatives to platonism. Weyl (1918, 15), for example, takes a "conceptualist stance" when explaining that extensions are completely different from the ordinary objects to which the corresponding concepts apply. Classes result by the "mathematical process" and belong to an entirely separate sphere of existence ("einer ganz andern Existenzsphäre"). Nominalists, as Paul Gilmore (cf. the contribution of Villadsen to the present volume) give a more mundane answer to the question: classes are just names of the predicates which are used to define them. For the nominalist, *Yellow is a colour* just means that *yellow* is a colour word (cf. Gilmore, 2001, 385), and comprehension just means to mention predicates instead of using them. The paradoxes, then, are due to missing care in observing the use-mention-distinction, (cf. Gilmore, 2001, 384).[14]

In his exposition of Frege's class theory, Geach (1955, 231f) gives quite another answer to the question for the "nature of classes". According to him, every odd object will do as an extenion. Furthermore, it is completely irrelevant for the sake of class theory what objects are actually chosen as extensions. Take, for instance, the concept $P(x)$ of being an undergraduate x taught logic by Geach in October 1952 (Geach, 1955, 232). We may chose a human being h as the extension $E = \{x|P(x)\}$ of this concept; thus $h := E$. Then, of course, we may ask whether h has been one of Geach's logic students in October 1952, thus whether $h \in E$, i.e., whether $h \in h$. We recognize that the (thus interpreted) logical conception of a class does not exclude unfounded classes. Thus we are once more back at the questions dealt with by Benz and Bab/Mahr/Wiecorek.

That the actual choice of extensions for concepts does not matter, means that not theorem of class theory hinges on this choice. Thus, according to Geach, these theorems are distinguished by being invariantly true under different choices. This seems to be an interesting consideration deserving further

---

[12] In an extension of his system, Ackermann (1965, 12) reintroduces a restricted form of extensionality for special predicates.

[13] Cp., for example, Bachmann (1935) for a constructivist reading of Frege.

[14] Besides constructivism, nominalism has been one of the earliest critics of both set and type theory. The interested reader is referred to the relevant writings of Stanisław Leśniewski and Leon Chwistek, *e.g.*, Leśniewski (1916) and Chwistek (1932/33), as well as to David Lewis's (1991) mereological re-construction of set theory.

investigations and seems to be somehow connected to Fregean ideas.[15] As regards Geach's example of a student being the extension of a concept which applies to himself, I would like to refer the reader to a remark of Frege in his epilogue to the GRUNDGESETZE DER ARITHMETIK, where he discusses Russell's paradox: "Von der Klasse der Menschen wird niemand behaupten wollen, dass sie ein Mensch sei [Nobody will assume of the class of all humans that it is human]"; Frege (1903, 253). Should (according to Frege) the assigment of extensions to concepts be not as arbitrary as Geach assumes?

By the example of this last question, the reader will surely recognize that the contributions to this collection lead up to interesting problems both on the level of historical and philosophical interpretation and on that of technical research.

References

Wilhelm Ackermann. Der Aufbau einer höheren Logik. *Archiv für mathematische Logik und Grundlagenforschung*, 7:5–22, 1965.

Peter Aczel. Frege structures and the notions of proposition, truth and set. In Jon Barwise, H. J. Keisler, and K. Kunen, editors, *The Kleene symposium*, pages 31–59. North-Holland, 1980.

Friedrich Bachmann. Frege als konstruktiver Logizist. Manuscript. Westfälische Wilhelms Universität. Münster, 1935. Published as an appendix to *Frege und die moderne Grundlagenforschung*. Ed. by Christian Thiel. Verlag Anton Hain, Meisenheim 1975, 160–168, 1935.

George Boolos. The iterative conception of set. *Journal of Philosophy*, 68:215–232, 1971. Reprinted: G. Boolos. *Logic, logic, and logic*. Harvard University Press, Cambridge MA, 1998.

Emile Borel. Cinq lettres sur la théorie des ensembles. *Bulletin de la Société de France*, 33:261–273, 1905.

Luitzen Egbertus Jan Brouwer. Die moeglichen Maechtigkeiten. In *Atti del IV Congresso Internazionale dei Matematici (Roma, 6-11 Aprile 1908)*, volume III, pages 569–571. Tipografia della R. Acdademia Lincei, Rome, 1909. Reprinted: Brouwer (1975, 102–104).

Luitzen Egbertus Jan Brouwer. Begründung der Mengenlehre unabhängig vom logischen Satz vom ausgeschlossenen Dritten. Erster Teil: Allgemeine Mengenlehre. *Verhandelingen der Koninklijke Akademie van wetenschappen te Amsterdam. 1^e sectie 12 no. 5*, 1918. Reprinted: Brouwer (1975, 150–190).

Luitzen Egbertus Jan Brouwer. Über die Bedeutung des Satzes vom ausgeschlossene Dritten in der Mathematik, insbesondere in der Funktionentheorie. *Journal für die reine und angewandte Mathematik*, 154:1–7, 1923. Reprinted: Brouwer (1975, 268–280). English translation: van Heijenoort (1967, 334–341).

---

[15] Cp. Schroeder-Heister (1987).

Luitzen Egbertus Jan Brouwer. *Collected works*, Vol. 1: *Philosophy and foundations of mathematics*. Edited by Arendt Heyting. North-Holland Publishing Company, 1975.

Cajetan. *Commentaria in Isagogen Porphyrii ad Praedicamenta Aristotelis*. Ed. curavit Maria-Hyacinthus Laurent. Institutum 'Angelicum', Rome, 1939. Original edition Pisa and Milano 1497.

Leon Chwistek. Die nominalistischen Grundlagen der Mathematik. *Erkenntnis*, 3:367–388, 1932/33.

Solomon Feferman. Non-extensional type-free theories of partial operations and classifications, i. In Justus Diller and Gert Heinz Müller, editors, $\models$ ISILC *Proof theory symposion. Dedicated to Kurt Schütte on the occasion of his 65th birthday*, pages 73–118. Springer, Berlin *etc.*, 1975.

Solomon Feferman. Categorical foundations and foundations of category theory. In Jaakko Hintikka and Robert E. Butts, editors, *Proceedings of the 5th International Congress of Logic, Methodology and Philosophy of Science*. Vol. 1: *Foundations of mathematics and computability theory*, pages 149–169. Reidel, Dordrecht, 1977.

Solomon Feferman. Toward useful type-free theories. *The journal of symbolic logic*, 49:75–111, 1982.

Abraham Adolf Fraenkel. *Zehn Vorlesungen über die Grundlegung der Mengenlehre*. Teubner, Leipzig, 1927. Reprint: Wissenschaftliche Buchgesellschaft, Darmstadt, 1972.

Gottlob Frege. Kritische Beleuchtung einiger Punkte in E. Schröders Vorlesungen über die Algebra der Logik. *Archiv für Philososohie. II. Abt. Neue Folge der Philosophischen Monatshefte: Archiv für systematische Philosophie*, 1:433–456, 1895. Reprinted: Gottlob Frege. *Kleine Schriften* Olms, Hildesheim and New York, 1990, 193–210.

Gottlob Frege. *Grundgesetze der Arithmetik begriffsschriftlich abgeleitet*, volume II. Pohle, Jena, 1903. Reprinted: Olms, Hildesheim, 1966.

Peter Thomas Geach. Class and concept. *Philosophical review*, 64:561–570, 1955. Reprinted: P. T. Geach. *Logic matters*. Oxford, Blackwell 1972, 226–235.

Paul Carl Gilmore. An intensional type theory: motivation and cut-elimination. *The journal of symbolic logic*, 66(1):383–400, 2001.

Hans Hermes and Heinrich Scholz. *Mathematische Logik*, volume I/1, Heft 1, Teil 1 of *Enzyklopädie der mathematischen Wissenschaften*. Teubner, Leipzig, 1952.

Stanisław Leśniewski. *Podstawi ogólnej teoryi mnogości, I*. Prace Polskiego Koła Naukowego w Moskwie. Sekcya matematyczno-przyrodnicza, no 2., Moscow, 1916. English translation *Foundations of the general theory of sets, I*. In Stanisław Leśniewsk. *Collected works, vol. I*. Kluwer, Dordrecht,1992, 129–173.

David Lewis. *Parts of classes*. Blackwell, Oxford, 1991.

Bernd Mahr. Modellieren. Beobachtungen und Gedanken zur Geschichte des Modellbegriffs. In Sybille Krämer and Horst Bredekamp, editors, *Bild — Schrift — Zahl*, pages 59 – 86. Fink, München, 2003.

Henri Poincaré. *Science et méthode*. Flammarion, Paris, 1908. English translation: *Science and method*. Routledge, London 1914. Reprinted 1996.

Peter Schroeder-Heister. A model-theoretic reconstructions of Frege's permutation argument. *Notre Dame journal of formal logic*, 28:69–79, 1987.

Stewart Shapiro, editor. *Intensional mathematics*. North-Holland Publishing Company, Amsterdam, 1985.

Jean van Heijenoort, editor. *From Frege to Gödel. A source book in mathematical logic, 1879–1931*. Harvard Universtiy Press, Cambridge MA, 1967.

John von Neumann. Eine Axiomatisierung der Mengenlehre. *Journal für die reine und angewandte Mathematik*, 154:219–240 (Berichtigung vol. 155, p. 128), 1925. English translation: van Heijenoort (1967, 393–413).

Hermann Weyl. *Das Kontinuum. Kritische Untersuchungen über die Grundlagen der Analysis*. Veit, 1918. Reprinted in *Das Kontinuum und andere Monographien*. Chelsea Publishing Company, New York (no date given). — English translation: *The continuum: a critical examination of the foundation of analysis*. Thomas Jefferson University Press, Kirksville MO, 1987.

Hermann Weyl. *Philosophie der Mathematik und Naturwissenschaft*. Oldenbourg, München and Wien, 3rd edition, 1966. English translation: *Philosophy of mathematics and natural science*. Princeton University Press, Princeton 1966.

Fixing Cantor's Paradise: The Prehistory of Ernst Zermelo's Axiomatization of Set Theory

Volker Peckhaus

Institute for Humanities: Philosophy
University of Paderborn

volker.peckhaus@upb.de

Summary. In 1925 David Hilbert appealed to his fellow mathematicians to do everything not to be driven out of Cantor's paradise. This appeal on the climax of the foundational crisis in mathematics had a long prehistory, released by problems of and open questions in set theory having become evident already in the end of the 19th century. Ernst Zermelo's axiomatization of set theory was a first step to overcome these problems within the framework of Hilbert's early axiomatic programme. In this early period the heavenly nature of Cantor's theory was not yet understood, but it nevertheless slowly grew into a cornerstone of the foundations of mathematics. The paper deals with the main motivations for Zermelo's axiomatization of set theory, in particular the impact of Russell's paradox which was independently found by Zermelo at almost the same time. It is argued that besides the paradoxes three aspects had a decisive influence: (1) Hilbert's axiomatic method, (ii) Cantor's Continuum Hypothesis, (iii) unintended sets in Cantor's theory. Zermelo's way to set theory is sketched and finally some hints on his later ideas on this topic connected to the foundational crisis in the 1920s are given.

1 Introduction

On June 4, 1925, the Göttingen mathematician David Hilbert took part in a meeting organized by the Westphalian Mathematical Society in Münster in honour of the memory of Karl Theodor Wilhelm Weierstraß. He delivered the address "On the Infinite" containing the famous exclamation coming close to an appeal: "No one shall be able to drive us from the paradise that Cantor created for us" (Hilbert, 1967, 376). He referred to the fact that set theory was threatened, and Hilbert gave the reasons: The paradoxes of set theory, "in particular, a contradiction discovered by Zermelo and Russell", which "had, when it became known, a downright catastrophic effect in the world of mathematics". According to Hilbert this effect led to "extremely vehement attacks" directed against Cantor's theory. "The reaction was so violent that the commonest and most fruitful notions and the very simplest and most important modes of inference in mathematics were threatened and their use was to be prohibited" (Hilbert, 1967, 375). This last remark was directed against the intuitionists at the climax of the struggle in foundational issues in mathematics.

In this paper "On the Infinite" Hilbert also gave systematic reasons for his high esteem of set theory. Analysis alone, he said, "does not yet give us the deepest insight into the nature of the infinite. Rather, this is conveyed to us only by a discipline that is closer to the general philosophical way of thinking and was destined to place the entire complex of questions concerning the infinite in a new light". And this was Cantor's set theory with its core, the theory of transfinite numbers. Hilbert sang the praise of this theory saying: "This appears to me to be the most admirable flower of the mathematical intellect and in general one of the highest achievements of purely rational human activity" (Hilbert, 1967, 373).

Hilbert's 1925 appeal to the mathematical community to do everything not to be driven out of Cantor's paradise should support his recent attempt to fix set theory in order to preserve it from developments seen as having been released by the discussion of the Zermelo-Russell paradox. This new attempt was Hilbert's meta-mathematics or proof theory designed as some sort of contentual pre-mathematics providing the proof techniques used in heavenly formalistic real mathematics.

My task is, however, not to speak about this later discussion of set theory in Göttingen, but on the early period in which the transition from Cantor's naive set theory to axiomatized set theory took place, in particular the prehistory of Ernst Zermelo's axiomatization of set theory presented in 1908. In the following I will speak about this early period, the time when the heavenly nature of Cantor's theory was not yet understood, but when it nevertheless slowly grew into a cornerstone of the foundations of mathematics. What were the main motivations for Zermelo's axiomatization of set theory? What was the impact of the paradoxes, in particular of Russell's paradox which was independently found by Zermelo at almost the same time. I will argue that three aspects had a decisive influence:

1. Hilbert's axiomatic method,
2. Cantor's Continuum Hypothesis,
3. unintended sets in Cantor's theory and the paradoxes.

I will then sketch Zermelo's way to set theory and finally close the circle by hinting at later developments connected to the foundational crisis in the 1920s.[1]

---

[1] An extensive presentation of these topics can be found in Ebbinghaus (2007).

2 The Prehistory of Zermelo's Axiomatization of Set Theory

2.1 Hilbert's Axiomatic Method

The date of birth of the modern axiomatic method is usually given with 1899, when David Hilbert's published his seminal *Grundlagen der Geometrie* (Hilbert, 1899), not really a book on method, but the application of a method, the axiomatic method, to Euclidean geometry.[2] Hilbert starts with definitions ("Erklärungen") proceeding from three imagined systems of things (points, straight lines, planes) which he called, using the Kantian term, "thought things", *i.e.*, empty concepts. He then stipulates their interrelations in a set of 20 axioms. He freed the axiomatic system from any intuition, so differing from Euclid's model. We should recall, Euclid started his *Elements* with definitions like "A point is that which has no part" or "A line is breadthless length". Such definitions require abstraction and, in order to make abstraction possible, an intuition of some sort of real world analogue of a point or a line.

Contrary to Euclid, Hilbert likewise justified his axiom system internally by investigating his set of axioms as an *object in itself*, proving its completeness, the independence of the axioms, and its consistency. In the case of Euclidean geometry the latter proof was done by reducing the consistency of the geometrical axioms to the presupposed consistency of arithmetic. Therefore, a complete consistency proof was in fact postponed, and, at the same time, a new task was set: to find a consistent set of axioms for arithmetic. Hilbert presented his ideas concerning the foundations of arithmetic in September 1899, at the annual meeting of the Deutsche Mathematiker-Vereinigung which took place at Munich. In his lecture "Über den Zahlbegriff" (Hilbert, 1900b) he elaborated the foundations of arithmetic — in his opinion the basic discipline of mathematics — independently of set-theoretic considerations. Due to the sketchy character of this paper, Hilbert did not carry out the meta-axiomatic investigations on independence of the axioms, completeness of the system and its consistency. Concerning the "necessary task" to prove consistency, he asserted, "only a suitable modification of known methods of inference" (Hilbert, 1900b, 184) was required. These optimistic words from September 1899 seem to indicate that Hilbert probably underestimated the enormity of the task in hand. He soon changed his views. In August 1900, less than one year later, he included the consistency proof for the axioms of arithmetic as the second among his famous mathematical problems which he presented to the Second International Congress of Mathematicians at Paris (cf. Hilbert, 1900a). Three years later, on 27 October 1903, he again emphasized the distinguished rôle of the consistency proof in a lecture delivered before the Göttingen Mathematical Society on the foundations of arithmetic. Following the report, it was Hilbert's aim "to work out the 'axiomatic' standpoint clearly". On the rôle of

---

[2] For the prehistory and development cf. Toepell (1999).

consistency, he then coined the brief formula: "the principle of contradiction the pièce de résistance."[3]

One should observe that Hilbert's early axiomatic method was not intended to serve as a universal way of presentation in mathematics. It was to be used in cases where the foundations of a theory were questioned. This was the case in Euclidean geometry, as Hilbert wrote to Frege,[4] where the status of the parallel axiom was still unclear. And this was the case in set theory, as well.

2.2 Continuum Hypothesis

The first of the problems Hilbert dealt with in his Paris lecture on "Mathematical Problems" was entitled "Cantor's Problem of the Cardinal Number of the Continuum", *i.e.*, the problem to give a direct proof of Cantor's continuum hypothesis which runs, in Hilbert's formulation, as follows (Hilbert, 1902b, 446):

> Every system of infinitely many real numbers, *i.e.*, every assemblage of numbers (or points), is either equivalent to the the assemblage of natural integers, 1, 2, 3, ... or to the assemblage of all real numbers and therefore to the continuum, that is, to the points of a line; *as regards equivalence there are, therefore, only two assemblages of numbers, the countable assemblage and the continuum.*

The required direct proof has not yet been found. Gödel proved in 1938 (Gödel 1938, 1939, 1940) that the continuum hypothesis was consistent to the customary axioms of set theory and much later Paul Cohen (Cohen 1963a, 1963b, 1964) that it was independent of these axioms.

In the Paris talk, Hilbert hinted at a "very remarkable statement of Cantor's" standing "in the closest connection with the theorem mentioned." He referred to Cantor's assumption that every set can be well-ordered. Hilbert conjectured, that the proof of this statement might offer the key to the proof of the continuum hypothesis.

2.3 Unintended Sets

In his correspondence with Georg Cantor between 1897 and 1900 Hilbert discussed problems of set theory between 1897 and 1900, above all those arising from the assumption of a set of all cardinals. Already in the first of Cantor's letters to Hilbert, dated 26 September 1897 (cf. Cantor, 1991, no. 156, pp. 388-389) Cantor proved that the totality of alephs does not exist, *i.e.*, that this totality is not a well-defined, finished set [*fertige Menge*]. If it is taken to be a finished set, a certain larger aleph would follow on this totality. So this

---
[3] Cf. the report in *Jahresbericht der Deutschen Mathematiker-Vereinigung* **12** (1903), 592.
[4] Cf. Hilbert's letter to Frege, dated 29 December 1899, Frege (1976), 66.

new aleph would at the same time belong to the totality of all alephs, and not belong to it, because of being larger than all alephs (*ibid.*, p. 388). Cantor consequently distinguished sets from other kinds of multiplicities, *i.e.*, "finished" sets from multiplicities which are no sets, like the totality of all cardinals. The latter multiplicities are "absolutely infinite", unlike the former ones, the "transfinite" sets. In a later letter Cantor gave the following characterization of a finished set: A set can be imagined as finished if it is consistently possible to imagine all of its elements as being gathered, the set itself therefore as *one* compound thing, *i.e.*, if it is possible to imagine the totality of its elements as existing.[5] This is, however, impossible for the absolute infinite which he identifies with God. Realized in its highest perfection in God it has to be strictly opposed to the *actual infinite* which he calls the transfinite (Cantor, 1887, 81-82).

Cantor disproves the existence of the totality of all cardinals by showing that the assumption of its existence contradicts his definition of a set as a comprehension of certain well distinguished objects of our intuition or our thinking in a whole.[6] The totality of all cardinals (and of all ordinals) cannot be thought of as *one* such thing, contrary to actual infinite objects like transfinite sets. He is therefore not really concerned with paradoxes and their solution, but with non-existence proofs using *reductio-ad-absurdum* arguments.[7]

It is clear that Cantor and Hilbert were concerned with what later was called "Cantor's paradox", *i.e.*, the paradox of the greatest cardinal, or of the set of all cardinals. It is also clear, however, that the contradiction discussed by Cantor served only as a paradigmatic example for other inconsistent multiplicities, *i.e.*, totalities resulting from unrestricted comprehension.

Hilbert's responses in correspondence have not been preserved, but he published his opinion at prominent places. As mentioned above, Hilbert gave a set of axioms for arithmetic in the paper "On the Concept of Number" from 1900, and claimed that only a suitable modification of known methods of inference would be needed for proving the consistency of the axioms. If this proof were successful, the existence of the totality of real numbers would be shown at the same time. In this context he referred to Cantor's problem as to whether the system of real numbers is a consistent, or finished, set. He stressed:

> Under the conception above, the doubts which have been raised against the existence of the totality of all real numbers (and against the existence of infinite sets in general) lose all justification; for by the set of real numbers

---

[5] Cantor's letter to Hilbert, dated 2 October 1897, Cantor (1991, 390).
[6] Cantor (1895), quoted in Cantor (1932, 282): "Unter einer 'Menge' verstehen wir jede Zusammenfassung M von bestimmten wohlunterschiedenen Objekten m unsrer Anschauung oder unseres Denkens (welche die 'Elemente' von M genannt werden) zu einem Ganzen."
[7] This opinion follows Moore and Garciadiego (1981), Garciadiego Dantan (1992).

> we do not have to imagine the totality of all possible laws according to which the elements of a fundamental sequence can proceed, but rather — as just described — a system of things whose internal relations are given by a *finite and closed* set of axioms [...], and about which new statements are valid only if one can derive them from the axioms by means of a finite number of logical inferences.[8]

He also claimed that the existence of the totality of all powers or of all Cantorian alephs could be disproved, *i.e.*, in Cantor's terminology, that the system of all powers is an inconsistent (not finished) set (ibid.). The condition was to reformulate set theory according to the "conception above", *i.e.*, to give a consistent axiomatic system from which Cantor's system could be derived deductively.

Hilbert took up this topic again in his Paris lecture on "Mathematical Problems".[9] In the context of his commentary on the second problem concerning the consistency of the arithmetical axioms he used the same examples from Cantorian set theory and the continuum problem as in the earlier lecture. "If contradictory attributes be assigned to a concept," he wrote, "I say, that mathematically the concept does not exist" (Hilbert, 1996b, 1105).

Hilbert expressed his conviction that a suitable axiomatization would be able to avoid the contradictions resulting from the attempt to understand absolute infinite multiplicities as units, because only those concepts had to be accepted which could be derived from an axiomatic base.

There is evidence that it was during his discussions with Cantor that Hilbert formulated a paradox of his own, later called in Göttingen "Hilbert's Paradox",[10] and first written down in the unpublished lecture course *Logische Principien des mathematischen Denkens* given to his Göttingen students in the summer-term of 1905.[11] Hilbert considered this paradox, resulting from the set-formation principles of union and self-mapping, as "purely mathematical" because he carefully avoided to use any concept from transfinite arithmetic. It is this paradox Hilbert referred to in his letter to Gottlob Frege of 7 November 1903 after having received the second volume of Frege *Grundgesetze der Arithmetik* (Frege, 1903) containing Freges admission that the logical system used there for the foundation of arithmetic had proved to be inconsistent. In this letter Hilbert commented on Frege's description of Russell's paradox in the postscript, and wrote that "this example" was already known in Göttingen. In a footnote he added "I believe Dr Zermelo discovered it three or four years ago after I had communicated my examples to him" and continued

> I found other even more convincing contradictions as long as four or five years ago; they led me to the conviction that traditional logic is inade-

---

[8] Hilbert (1996b, 1095). German original Hilbert (1900a, 184).
[9] Hilbert (1900b), English translations Hilbert (1902b), Hilbert (1996a).
[10] Cf. Peckhaus and Kahle (2002).
[11] For an analysis of the lecture course cf. Peckhaus (1990a, ch. 3).

quate and that the theory of concept formation needs to be sharpened and refined.[12]

In this passage Ernst Zermelo is mentioned as having found Russell's Paradox already before the turn of the century. Russell's paradox was first published only in 1903 when Russell devoted chapter X of his *Principles of Mathematics* (Russell, 1903) to "The contradiction" and Frege had to admit in the epilogue of the second volume of his *Grundgesetze der Arithmetic* that the logical system presented there had been proved to be inconsistent. Nevertheless, Hilbert's priority claim can partially be verified. Among the papers of Edmund Husserl, until 1916 professor of philosophy in Göttingen, a note from Husserl's hand was found, in parts written in Gabelsberger shorthand, saying that Zermelo had informed him on 16 April 1902 that the assumption of a set M that contains all of its subsets $m, m', \ldots$ as elements, is an inconsistent set, *i.e.*, a set which, if treated as a set at all, leads to contradictions.[13]

I would like to stress again that Hilbert and Zermelo did not publish their paradoxes. Obviously they were not alarmed to a degree we would expect today. They were confident to have the tool at hand to get rid of the contradictions: a suitable axiomatization of set theory. Such axiomatization required a consistency proof for the axioms of set theory, and only the problems in finding such a proof led to the insight into the consequences of the Zermelo-Russell Paradox. Frege's logical system of the *Grundgesetze der Arithmetic* was affected and with it the most elaborated system of logic. As became clear, it made no sense to prove the consistency of arithmetic with the help of a logic just having been proved to be inconsistent.

3 Zermelo

Let's turn to Zermelo's part in this story. Zermelo came to Göttingen in 1897 in order to work for his *Habilitation*. His special field of competence was applied mathematics and mathematical physics, in particular the calculus of variations, thermodynamics and hydrodynamics.[14] After Hilbert's great Paris problems lecture he slowly changed the focus of his interests to set theory and foundations. He became Hilbert's first collaborator in the philosophy of mathematics, a first member of Hilbert's school before it was established. Already in

---

[12] Frege (1980, 51). German original (Frege, 1976, 79-80).
[13] Critical edition in *Husserliana* XXII (Husserl, 1979, 399): "*Zermelo* teilt mit (16. April 1902) [...] Eine Menge M, welche *jede* ihrer Teilmengen m, m' ... *als Element* enthält, ist eine inkonsistente Menge, d. h. eine solche Menge, wenn sie überhaupt als Menge behandelt wird, führt zu Widersprüchen." English translation in Rang and Thomas (1981).
[14] On Zermelo's activities in Göttingen cf. esp. Moore (1982), Peckhaus (1990b, 76-122), Peckhaus (1990a), Ebbinghaus (2007).

the winter semester 1900/01 he gave a first course on set theory in Göttingen.[15] In 1902 he published a first paper on the addition of transfinite cardinals (Zermelo, 1902). In August 1904 he was involved in the discussion of Julius König's rejection of the continuum hypothesis at the Third International Congress of Mathematicians in Heidelberg. One month later he communicated in a letter to Hilbert that he was able to prove the well-ordering theorem which Hilbert had named "the key for proving the continuum hypothesis" in the Paris problems lecture. The proof was published in the *Mathematische Annalen* almost immediately (Zermelo, 1904), keeping the letter form. It evoked a storm of protest in the mathematical community, above all directed against Zermelo's use of the principle of choice (cf. Moore, 1982). He reacted by publishing a new proof in 1908 together with a rejection of the main criticisms which was in parts very polemical (Zermelo, 1908b). In the same year Zermelo published an axiom system of set theory according to Hilbert's model (Zermelo, 1908a), sharing Hilbert's opinion that a "deepening of foundations" of a mathematical discipline was necessary as soon as this discipline was questioned because of foundational problems. This situation was given for set theory because of the paradoxes as was well known in Göttingen, at least in 1908.

Zermelo followed Hilbert's suggestions concerning the structure of axiomatic systems. Zermelo only mentions, however, the necessity of proving the consistency of his axioms. He did not present this proof, although he had intended to add such a proof as becomes evident from his correspondence with Hilbert. He stopped the research because Hilbert urged him to publish his results, a wise decision.[16]

4 Later Development

The further development is given in a telegraphic style: In 1908 Zermelo got a paid lectureship for "Mathematical Logic and Related Subjects", the first official lectureship for mathematical logic in Germany.[17] This indicates a shift in interest in Göttingen mathematics towards research in logic which became plain after the First World War, when the system of Alfred North Whitehead's and Bertrand Russell's *Principia Mathematica*, Vol. I (1910) became the basis for developing a logical system of their own as finally presented in Hilbert's and Ackermann's *Grundzüge der Theoretischen Logik* (1928). Zermelo had, however, problems to get a professorship mostly due to his tuberculosis which forced him to spend much time in Swiss sanatoriums.

---

[15] Cf. the notebook with the manuscript of this course (partially in shorthand) in the Zermelo papers, University Archives (UA), Freiburg i. Br., C 129/150. For a discussion cf. Moore (1982, 155-156).

[16] On Zermelo's attempts to proof the consistency of his axiom system, cf. his letter to Hilbert, dated Arosa, 25 March 1907, Hilbert papers, Staats- und Universitätsbibliothek Göttingen, Cod. Ms D. Hilbert 447, fol. 5.

[17] Cf. Peckhaus (1992).

In 1910 he received a call as full professor for mathematics at the University of Zurich, a position which he had to give up in 1916 due to his illness. From then on, he had to live from his Swiss pension. He moved to the Black Forest in Germany where he was able to establish contact to the mathematicians in Freiburg. From 1926 until his dismission in 1935 he had a honorary professorship for mathematics at the University of Freiburg which was not connected to revenues. During this time he began to publish again not only in the calculus of variations (game theory, rating system for strategy games, navigation problem in aviation mathematics), but also in set theory, thereby taking position on the climax of the foundational struggle in mathematics.

The most important of his contributions was, no doubt, his paper "On Boundary Numbers and Domains of Sets: New Investigations in the Foundations of Set Theory", published in 1930, which he described in a letter to the Warsaw mathematician Władysław Sierpiński[18] as dealing with the foundations of set theory and promising

> to give a satisfactory clarification of the so-called antinomies [...]. It concerns the investigation of such "domains of sets", in which the general axioms of set theory are satisfied, and the systematic development of their, essentially different (not isomorphic) "models" which can serve as their representations.

Its essential result is, as Geoffrey Hellman put it: "Set theory should be seen, not as the theory of a unique, all-embracing structure, but instead as a theory of an endless infinity of intimately related structures" (Hellman, 1989, 55–56).

Zermelo's paper on boundary numbers belongs to the context of his fight against, what he called, "Skolemism,"[19] i.e., any kind of finitism. It is interesting to see that Zermelo opposed almost all foundational positions at that time. He particularly did not follow Hilbert's move towards metamathematics. In metamathematics a constructive or operative way of founding mathematics is proposed which comes close to Brouwer's intuitionism in its restriction to finite operations. Zermelo, however, rejected any finitistic approach to mathematics as expression of a "Skolemism" in set theory. Alternatively he attempted to justify his infinite hierarchies of sets with the help of what he called a "logic of the infinite" (Zermelo 1932b, 1932a), trying to counter the "shortcomings of any 'finitistic' proof theory" (Zermelo, 1932b, 87). With this he became one of the early precursors of modern infinitary logic (Moore, 1997), and contributed, as the founder of standard set theory to alternative set theories.

References

Georg Cantor. Mitteilungen zur Lehre vom Transfiniten. *Zeitschrift für Philosophie und philosophische Kritik*, 91:81–125, 1887. Reprinted in Cantor

---

[18] Zermelo to Sierpiński, dated Freiburg, 26 March 1930, UA Freiburg, C 129/100.
[19] This term is used in Zermelo (1932b, 85).

(1932, 378–419).
Georg Cantor. Beiträge zur Begründung der transfiniten Mengenlehre (Erster Artikel). *Mathematische Annalen*, 46:481–512, 1895. Reprinted in Cantor (1932, 282–311).
Georg Cantor. *Gesammelte Abhandlungen mathematischen und philosophischen Inhalts. Mit erläuternden Anmerkungen sowie mit Ergänzungen aus dem Briefwechsel Cantor–Dedekind*. Ed. by Ernst Zermelo. Springer, Berlin, 1932. Reprinted: Olms: Hildesheim 1962, 1980; Springer: Berlin *etc.* 1980.
Georg Cantor. *Briefe*. Ed. Herbert Meschkowski and Winfried Nilson. Springer, Berlin *et al.*, 1991.
Paul Cohen. The independence of the continuum hypothesis. Mimeographed, Stanford University, 1963a.
Paul Cohen. The independence of the continuum hypothesis. *Proceedings of the National Academy of Science of the U.S.A*, 50:1143–1148, 1963b.
Paul Cohen. The independence of the continuum hypothesis II. *Proccedings of the National Academy of Science of the U.S.A*, 51:105–110, 1964.
Heinz-Dieter Ebbinghaus (in collaboration with Volker Peckhaus). *Ernst Zermelo. An approach to his life and work*. Springer, Berlin *et al.*, 2007.
William Ewald, editor. *From Kant to Hilbert: A source book in the foundations of mathematics, 2 Vols*. Clarendon Press, Oxford, 1996.
Gottlob Frege. *Grundgesetze der Arithmetik begriffsschriftlich abgeleitet*, Vol. 2. Hermman Pohle, Jena, 1903.
Gottlob Frege. *Wissenschaftlicher Briefwechsel*, ed. G. Gabriel *et al.* Meiner, Hamburg, 1976.
Gottlob Frege. *Philosophical and mathematical correspondence*, ed. by G. Gabriel *et al.* Basil Blackwell, Oxford, 1980.
Alejandro J. Garciadiego Dantan. *Bertrand Russell and the set-theoretic "paradoxes"*. Birkhäuser, Basel, Boston, and Berlin, 1992.
Kurt Gödel. The consistency of the axiom of choice and of the generalized continuum hypothesis. *Proceedings of the National Acadamy of Science of the U.S.A*, 24:56–557, 1938.
Kurt Gödel. Consistency proof for the generalized continuum-hypothesis. *Proceedings of the National Academy of Science*, 25:220–224, 1939.
Kurt Gödel. *The consistency of the continuum hypothesis*. Princeton University Press, Princeton, N.J., 1940.
Jean van Heijenoort, editor. *From Frege to Gödel. A source book in mathematical logic*. Harvard University Press, Cambridge, Mass., 1967.
Geoffrey Hellman. *Mathematics without numbers. Towards a modal-structual interpretation*. Clarendon Press, Oxford, 1989.
David Hilbert. *Grundlagen der Geometrie*. Teubner, Leipzig, 1899. 14th ed. Hilbert (1999), English translation Hilbert (1902a).
David Hilbert. Mathematische Probleme. Vortrag, gehalten auf dem internationalen Mathematiker-Kongreß zu Paris 1900. *Nachrichten von der königlichen Gesellschaft der Wissenschaften zu Göttingen, Mathematisch-*

physikalische Klasse aus dem Jahre 1900, pages 253–297, 1900a. English translation Hilbert (1902b).

David Hilbert. Über den Zahlbegriff. *Jahresbericht der Deutschen Mathematiker-Vereinigung*, 8:180–184, 1900b. English translation Hilbert (1996b).

David Hilbert. *Foundations of geometry*. Open Court, Chicago, 1902a. English translation of Hilbert (1899).

David Hilbert. Mathematical problems. Lecture delivered before the International Congress of Mathematicians in Paris 1900. *Bulletin of the American Mathematical Society*, 8:437–479, 1902b. English translation of Hilbert (1900a).

David Hilbert. Über das Unendliche. *Mathematische Annalen*, 95:161–190, 1926. English translation Hilbert (1967).

David Hilbert. On the infinite. In Heijenoort (1967, 367–392). Harvard University Press, 1967. English translation of Hilbert (1926).

David Hilbert. From "mathematical problems". In Ewald (1996, Vol. 2, 1096–1105). Oxford University Press, 1996a. Partial reprint of Hilbert (1902b).

David Hilbert. On the concept of number. In Ewald (1996, Vol. 2, 1089–1095). Harvard University Press, 1996b. English translation of Hilbert (1900b).

David Hilbert. *Grundlagen der Geometrie. Mit Supplementen von Paul Bernays, ed. by Michael Toepell*. Teubner, Stuttgart und Leipzig, 1999. 14th edition of Hilbert (1899).

David Hilbert and Wilhelm Ackermann. *Grundzüge der theoretischen Logik*. Springer, Berlin, 1928.

Edmund Husserl. *Aufsätze und Rezensionen (1890–1910), mit ergänzenden Texten*. Husserliana, Vol. XXII, ed. B. Rang. Nijhoff, The Hague, Boston, and London, 1979.

Gregory H. Moore. *Zermelo's axiom of choice*. Studies in the history of mathematics and physical sciences 8. Springer, Berlin, Heidelberg, and New York, 1982.

Gregory H. Moore. The prehistory of infinitary logic: 1885–1955. In M. L. Dalla Chiara *et al.*, editor, *Structures and norms in science*, pages 105–123. Kluwer Academic Publishers, 1997.

Gregory H. Moore and Alejandro Garciadiego. Burali-Forti's paradox: a reappraisal of its origin. *Historia Mathematica*, 8:319–350, 1981.

Volker Peckhaus. *Hilbertprogramm und Kritische Philosophie. Das Göttinger Modell interdisziplinärer Zusammenarbeit zwischen Mathematik und Philosophie*. Studien zurWissenschafts-, Sozial- und Bildungsgeschichte der Mathematik 7. Vandenhoeck & Ruprecht, Göttingen, 1990a.

Volker Peckhaus. Ich habe mich wohl gehütet, alle Patronen auf einmal zu verschießen. Ernst Zermelo in Göttingen. *History and philosophy of logic*, 11:19–58, 1990b.

Volker Peckhaus. Hilbert, Zermelo und die Institutionalisierung der mathematischen Logik in Deutschland. *Berichte zur Wissenschaftsgeschichte*, 15:27–38, 1992.

Volker Peckhaus and Reinhard Kahle. Hilbert's Paradox. *Historia Mathematica*, 29:157–175, 2002.

Bernhard Rang and Wolfgang Thomas. Zermelo's discovery of Russell's paradox. *Historia Mathematica*, 8:15–22, 1981.

Bertrand Russell. *Principles of mathematics*, volume 1. The University Press, Cambridge, 1903.

Michael Toepell. Zur Entstehung und Weiterentwicklung von David Hilberts 'Grundlagen der Geometrie'. In (Hilbert, 1999, 283–324). Teubner, 1999.

Alfred North Whitehead and Bertrand Russell. *Principia Mathematica*. Cambridge University Press, 1910. Vol. II (1912), Vol. III (1913), 2nd edition 1925–1927.

Ernst Zermelo. Ueber die Addition transfiniter Cardinalzahlen. *Nachrichten von der Königl. Gesellschaft der Wissenschaften zu Göttingen. Mathematisch-physikalische Klasse aus dem Jahre 1901*, pages 34–38, 1902.

Ernst Zermelo. Beweis, daß jede Menge wohlgeordnet werden kann. *Mathematische Annalen*, 59:514–516, 1904. English translation Zermelo (1967b).

Ernst Zermelo. Neuer Beweis für die Möglichkeit einer Wohlordnung. *Mathematische Annalen*, 65:107–128, 1908a.

Ernst Zermelo. Untersuchungen über die Grundlagen der Mengenlehre. i. *Mathematisch Annalen*, 65(261–281), 1908b. English translation Zermelo (1967a).

Ernst Zermelo. Über mathematische Systeme und die Logik des Unendlichen. *Forschungen und Fortschritte*, 8:6–7, 1932a.

Ernst Zermelo. Über Stufen der Quantifikation und die Logik des Unendlichen. *Jahresbericht der Deutschen Mathematiker-Vereinigung*, 41:2nd sect., 85–88, 1932b.

Ernst Zermelo. Investigations in the foundations of set theory i. In (Heijenoort, 1967, 199–215). English translation of Zermelo (1908a), 1967a.

Ernst Zermelo. Proof that every set can be well-ordered. In Heijenoort (1967, 139–141). Harvard University Press, 1967b. English translation of Zermelo (1904).

Ackermann's Class Theory

Klaus Robering

Institute of Business Communication and Information Science
University of Southern Denmark

robering@sitkom.sdu.dk

Summary. This article deals with Ackermann's attempt to build up a provably consistent system of class theory which allows for unrestricted comprehension. Ackermann developed several such systems, which all are based on two basic ideas. The first is to switch to a logic allowing for truth value gaps. The second idea involves a special interpretation of the conditional connective. $A \to B$ is interpreted as stating that B may be inferred from A. In this article, we apply a special variant of sequent logic invented by Ackermann for the motivation of his logic of entailment in order to analyse the special conditional connective of his class theory. The article starts with a presentation of Ackermann's analysis of the paradoxes of class theory. Then the logic underlying one of Ackermann's systems of class theory is developed and put into a broader perspective by comparing it to other logics. Finally, we briefly consider the fate of the paradoxes of Russell and Curry within Ackermann's system. The whole article is a contribution to what may be called "the proof theory of classes".

1 Introduction

Wilhelm Ackermann (1896–1962) is well known for his many contributions to various branches of logic such as proof theory, decidability, recursive function theory, axiomatic set theory, and relevance logic. This article deals with a far less known part of Ackermann's work, namely with his attempts to build up a provably consistent system of logic which could serve as a foundation of mathematics but would be free both from the restrictions of type theory and that of axiomatic set theory. For Ackermann, one essential criterion for a system's usefulness as a foundational framework is the possibility to formulate arithmetic and analysis, thus the theories of the natural and real numbers, in it. Ackermann's first article on these matters appeared 1941 and his last article, posthumously published in 1965, is still devoted to this program; cf. Ackermann (1941, 1965). The development of number theory and analysis in these systems is treated in Ackermann (1953, 1959, 1965). Approximately a third of Ackermann's publications is concerned with the problem of a type free logic.[1]

Feferman (1982, 106) remarks that Ackermann's systems "... are not keyed to any prior semantics". This is certainly true when "semantics" is taken in the model-theoretic sense. However, Ackermann's systems rest on two very clear ideas. The first of these is inspired by Heinrich Behmann's analysis of the paradoxes, see Behmann (1931a); it is to use some kind of partial logic. The second idea is an outcome from Ackermann's analysis of Curry's paradox; cp. Curry (1942). Ackermann concludes from this paradox that it is not sufficient to shift towards a partial logic in order to avoid the paradoxes but that the treatment of the conditional connective has to be changed, too. In his articles he suggest two strategies for such a change.

Here we shall first explain the contentual conceptions and ideas unerlying Ackermann's systems.[2] In the second part of the article, we shall re-formulate Ackermann's class theory using a method developed by him for motivating his logic of entailment. This method, namely the employment of Σ-sequents, highlights the essential idea in Ackermann's suggestion and at the same time makes transparent the relation of Ackermann's system to other logics. In the final section we loook on the fate of the paradoxes of Russell and Curry in Ackermann's class theory.

2 The Theory of Classes

2.1 Two Concepts of Classes

Basic concepts tend to have intricate histories, and this is especially true for such notions as 'set' and 'class'. There are several, not always compatible traditions leading up to the formation of set theory in the last third of the 19th and the first third of the 20th century. In philosophy and logic, classes figured as extensions of concepts, and this tradition reaches back to the logical theories of. Aristotle[3] On the other hand, mathematicians conceived of sets as

---

[1] A brief biography of Ackermann containing a bibliography of his writings can be found in Hermes (1967).

[2] The reader should be aware that these systems do not include what is known as "Ackermann's set theory", *i.e.*, the system developed in Ackermann (1956b). Applying the distinction explained in subsection 2.1. below, that system — as its name rightly suggests — is a system of *set* theory whereas the systems dealt with in the present article are most appropriately conceived of as explications of the *logical notion of a class*. Reinhardt (1970) has proved, that Ackermann's set theory is, in a sense, equivalent to ZF. Thus it is based on the iterative conception of a set which differs from the logical conception of a class as an extension of a concept. — One of the ideas of Ackermann's set theory is implemented in Grue's map theory; cf. the present volume p. 104.

[3] Cf., for instance, Walther-Klaus (1987) for a history of the traditional doctrine of concepts, their extensions and contents.

'aggregates' composed in some way of its elements, and this view roots in such purely mathematical endeavours as, for instance, combinatorics, probability theory, geometry, and analysis; see, *e.g.*, (Tiles, 1989, chps. 3, 4). Corresponding to the two traditions mentioned, there are two quite different concepts of a class. As already hinted at, the 'collective'[4] conception takes classes as composed of its members; they are collections of their members. In contrast to this, the so-called 'logical' conception conceives of classes as objects representing properties. In the following, I will reserve the label *set* for collective approaches; and I will call these approaches *set theories*. The words *class* and *class theory*, however, I will reserve for the "logical" approach.

Cantor's famous explanation of a set from (1895/1897) as a "Zusammenfassung M von bestimmten wohlunterschiedenen Objekten m unserer Anschauung oder unseres Denkens (welche die 'Elemente' von M genannt werden) zu einem Ganzen" clearly belongs to the collective tradition and thus marks a starting point of set theory. Furthermore, Frege whith his doctrine of classes as 'Werthverläufe' (courses-of-values), as extensions of concepts is a typical representative of the "logical" conception and is a starting point — not of class-theory in itself — but of formalized class-theory. Of course, representing in the class theorist's manner concepts by classes (as their extensions) provokes the question for what purpose at all we need such a representation. Can we not just rest content with the concepts itselves? Part of the answer is that classes do not come alone; they are accompanied by the relation $\in$ of membership. This relation enables us to simulate the predicative behaviour of properties by means of their corresponding classes: instead of saying that E applies to a we just say that $a \in \{x|Ex\}$, *i.e.*: that a is a member of the class of all Es. Thus we represent properties by classes in the way shown by the diagram of Fig. 1. The class $\{x|Ex\}$ is an individual as its possible element a. Class and element are items on the same level whereas element and concept belong to different logical levels. Now suppose you want to express that a has some property of a certain kind. Without classes, you have to express this by some second-order expression like this: "$\exists P.[Pa \wedge \ldots]$". Having classes available, however, you can stay first-order by writing: "$\exists x.[a \in x \wedge \ldots]$.

2.2 The Zermelo-Russell Paradox

Frege, as I said above, marks the beginning of formal class theory. Since he made his basic assumptions completely transparent by putting forward a formal system, it could be recognized that something was wrong with them and,

---

[4] The term *collective* is used by Leśniewski, for example in Leśniewski (1931, Fn. 10), in a very special technical sense relating to his theory of the part-whole-relationship. Here, however, it is used in a more informal way to denote all approaches — including the so-called 'iterative or cumulative conception of sets' — which conceive of sets as composed in some way of their members. This does not necessarily imply that the members are parts of the classes in the Leśniewskian understanding of that term.

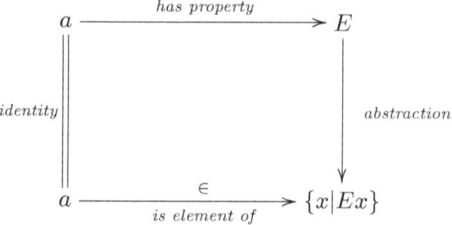

Fig. 1. Predication and Membership

more generally, that there is a problem with class theory. Frege wanted to represent concepts by objects in the way shown in figure 1 above. Obviously then, concepts applying exactly to the same objects have to be represented by the same object. From the set theoretic viewpoint it is readily seen that this kind of representation is impossible: Let D be the realm of all objects. Now we look at the properties defined on D. Working within set theory, we model them by the subsets of D. The totality of properties, then, is just the powerset $\wp(D)$ of D. What is tried in the figure above is to define a one-one mapping between $\wp(D)$ and D. As Cantor, employing his famous diagonal argument, found out already before 1890, this is impossible, since $\wp(D)$ exceeds D in cardinality. Frege actually reached the same result — though *via* the hard way.

Set theory is by no means struck by the paradoxes in the same way as class theories. Sets, according to the basic tenet of that approach, are collections; and there are no *a priori* reasons to believe that every odd totality of object may be assembled into one single coherent whole. Already Cantor distinguishes between totalities whose members can be united within a set and "inconsistent" totalities where this is impossible.[5] The late John Myhill (1984, p. 129) reports that

> Gödel said to me more than once, "there never were any set-theoretic paradoxes, but the property-theoretic paradoxes are still unresolved", and he may well have said the same thing in print.[6]

The principles of a set theory which state which manifolds of elements make up a set are called "comprehension principles". As just stated, there is no *a*

---

[5] "Eine Vielheit kann nämlich so beschaffen sein, daß die Annahme eines 'Zusammenseins' *aller* ihrer Elemente auf einen Widerspruch führt, so daß es unmöglich ist, die Vielheit als eine Einheit, als 'ein fertiges Ding' aufzufassen. Solche Vielheiten nenne ich *absolut unendliche* oder *inkonsistente Vielheiten*; (Cantor, 1932, 443)". Cf. Peckhaus's explanation of the role of "unintended sets" for the development of axiomatic set theory; this volume, page 14.

[6] Indeed, Gödel did so. In his (1947, p. 518), he says about the paradoxes: "They are a very serious problem, but not for Cantor's set theory". Cf. on this matters Peckhaus's contribution to this volume, esp. p. 15.

priori reason to accept the unrestricted comprehension principle according to which each manifold determines a set. On the contrary, the paradoxes show that the unrestricted principle is to be rejected. On the other hand, the principle that every concept has an extension only concerns the logical concepts of property and extension. It seems to be a principle which has been accepted without serious questioning and was shaken only by the paradoxes. It is with regard to these logical matters that Russell brought "to light the amazing fact that our logical intuitions (*i.e.*, intuitions concerning such notions as: truth, concept, being, class, *etc.*) are self-contradictory"; (Gödel, 1944). Furthermore, the axiomatic restrictions dictated, *e.g.*, by Russell's doctrine of types or by Zermelo's cumulative hierarchy, which banish (as far as we know) the pradoxes, are completely arbitrary when applied to the logical concept of a class.

In the framework of first-order logic, properties may be represented by open formulas. Unrestricted comprehension, then, is naturally implemented by adopting a term building operator converting property expressions, thus open formulas, into terms. Applying the terminology of combinatory logic, Ackermann (1958, 3) calls a system of logic "combinatorially complete" if each property expressed by a formula "$A[a]$" with a free variable "a" determines a corresponding class denoted by "$\{x|A[x]\}$" such that $a \in \{x|A[x]\}$ iff $A[a]$. Adopting the strategy of set theory amounts to acknowledging that some of the terms $\{x|A[x]\}$ lack denotation. In order to avoid complications in the underlying logic, the best, then, would be to give up combinatory completeness and to abandon the term building operator $\{x| - - - x - - -\}$. If, however, one nevertheless wants to keep the very idea of class theory, retaining combinatorially completeness is still, in spite of the paradoxes, the most reasonable starting point. Adopting unrestricted class comprehension as a logical principle forces then an adjustment of other logical principles.

3 Behmann's Diagnosis of the Paraoxes

One principled way for such an adjustment has been proposed by Heinrich Behmann (1931a). Behmann's analysis has been Ackermann's main source of inspiration. Behmann observes that Russell's paradox arises when one tries to find out whether the Russellian set $R \underset{\text{def}}{=} \{x | x \notin x\}$ is an element of itself or not. Both possibilities lead up to a contradiction, hence the paradox. Obviously Behmann is thinking of an argument by "classical dilemma", the two possibilities $R \in R$ and $R \notin R$ being the horns of the dilemma. An obvious way out is to reject the instance $R \in R \vee R \notin R$ of the *tertium non datur* by assuming that the $\in$-relation is only partially defined and that this relation fails short of determining a truth-value for the pair (R, R).[7]

---

[7] This strategy has also been applied (independently of Behmann), for instance, by Church (1932), who besides Behmann seems to have been another source of in-

Behmann, however, does not rest content in declaring membership to be a partial relation. He tries to find an explanation for this partiality and finds it in the fact that the abstraction term R cannot be eliminated from the formulas used in the derivation of Russell's paradox. Abstraction and membersphip are tied together by the principle

$$a \in \{x|A[x]\} \leftrightarrow A[a], \tag{1}$$

which obviously is a conversion principle corresponding to β-conversion in λ-calculus. What Behmann observes is that a formula like $R \in R$ is not "normalizable"; cf. Grues contribution to this volume, p. 90. Indeed, applying (1) to $R \in R$ one gets[8] $\sim R \in R$, from which $\sim\sim R \in R$, thus $R \in R$, results, *etc*. Now, (1) is interpreted by Behmann as a definition explaining the use of the abstraction term. For Behmann then, the fact that we cannot eliminate the sign "R" from "$R \in R$" is just a violation of the requirement that defined terms should be eliminable. Such cases of non-eliminability give rise to defintional gaps of predicates.

Ackermann, Bernays, and Gödel objected[9] to this and pointed out that it is also possible to derive, *e.g.*, Russell's paradox without making use of abstraction terms. However, Behmann in his reply (1931b) reinforces a point already to be found in his original article (1931a), namely that the admittance of partial predicates makes necessary a revision of quantification theory and that the logical principles used in the derivations at issue are no more valid in his revised logic. In his first attempt towards a typefree system of logic, Ackermann (1941, 8) admits that Behmann is right in his defense. On the other hand, however, he conjectures that Behmann's suggestions towards a modification of predicate logic might well be too restrictive for building up a useful system able to meet the requirements of mathematics (see Ackermann, 1941, 9).

3.1 The Conditional and Curry's Paradox

A logic admitting for partial predicates imediately rises a problem for the classical treatment of the conditional connective. Let P be a predicate with

---

spiration for Ackermann. Later the same idea (again independently of Behmann) was used by Fitch (1948). In a whole series of papers, Fitch made it the basis for his reconstruction of classical analysis. One version of his system of logic is described in his monography (1952). Halldén (1949), in building up his "logic of nonsense", takes Behmann's ideas about partially defined predicates as a starting point for his discussion. Both the work of Fitch and that of Halldén has been reviewed by Ackermann; see his (1950a) and (1952a).

[8] The sign "$\sim$" is used for the strong negation operation which interchanges the truth values. Negation defined in terms of absurdity, see (12) below, is denoted by "$\neg$".

[9] On these matters compare Thiel (2002).

"gaps"; then it is natural to reject $\sim P(a) \vee P(a)$ since the formula $P(a)$ is neither true nor false for some values of the parameter a. But surely, we would like to keep the theorem $\forall x.[P(x) \to P(x)]$, which classically is equivalent to $\forall x.[\sim P(x) \vee P(x)]$, the generalization of a formula we just rejected. The possibility of equating $A \to B$ with $\sim A \vee B$ is thus lost in a system of partial logic and the question arises how to characterize adequately the conditional connective within such a system.

When Ackermann (1950b), after his first attempt toward a system of unrestricted class theory in his (1941), returns to these matters, his discussion centers around the problem of the conditional. His aim in (1941) has not only been just to set up a system of typefree logic, his ambition was much more far reaching, namely to prove by metamathematical means the consistency of that system. Ackermann (1950b, 33) admits that his attempts in this directions had failed: he was only able to prove his original system consistent when giving up two axioms, whose purpose has been to ensure the validity of a deduction theorem. In his following articles from the 1950s, he emphasizes this issue pointing out in this connection the importance of Curry's paradox (see Curry, 1942). Using only logical principles involving the conditional, Curry's paradox demonstrates that a combinatorically complete system is inconsistent in the Post sense, i.e., every formula A is derivable in such a system. Let C_A (Curry's class with respect to A) be $\{x | x \in x \to A\}$; then we argue as follows.[10]

(1) $C_A \in C_A$ assumption   (2)
(2) $C_A \in C_A \leftrightarrow [C_A \in C_A \to A]$ instance of principle 1
(3) A (1), (2), *modus ponens* twice
(4) $C_A \in C_A \to A$ (1), (3), deduction theorem
(5) $C_A \in C_A$ (4), (2), *modus ponens*
(6) A (5), (2), *modus ponens* twice

A system of logic for which a deduction theorem holds true is called "deductively complete" by Ackermann, presumably because in such a system each deduction $A \vdash B$ is represented by a corresponding theorem $\vdash A \to B$. Thus the system completely mirrors all its derivations by corresponding theorems. In his (1958, 3) then, Ackermann formulates the basic problem concerning class theory in the following way: "Eine Logik kann nicht zugleich kombinatorisch und deduktiv vollständig sein, wenn die Widerspruchsfreiheit gewahrt werden soll".

3.2 Ackermann's Systems

A class theory with unrestricted comprehension is thus forced to give up the deduction theorem. Of course, one can try to develop a comprehensive system

---

[10] In the following deduction, the occurence of A in line (3) depends on the assumption of line (1), of course. In contrast, the occurence of the same formula in line (6) is not dependendent on any hypothesis.

of logic without a conditional connective. If one wants to retain it, however, the question arises how to treat it formally. Ackermann in his typefree systems applies two different strategies as regards the conditional. According to the respective strategy applied, we may thus classify his systems as belonging to two families: the first comprising the systems of Ackermann (1941, 1958, 1959), the second family including Ackermann (1950b, 1952b, 1965).

The most important use of the conditional is to express, in collaboration with the universal quantifier, subsumption of one concept under another: $\forall x.[F(x) \to G(x)]$. Ackermann's systems of the first kind exclusively take care of this contexts of use and lack a conditional connective for statements. For expressing subsumption in these systems, Ackermann employs what we nowadays might call a family of generalized binary quantifiers. The subsumption of F under G would be expressed by $F(x) \to_x G(x)$; that R is a binary subrelation of Q by $R(x,y) \to_{xy} Q(x,y)$, etc. In contrast to this, the systems of the second family retain the conditional and universal quantification as two seperate notions. This gives these systems a much more standard appearance than those of the first family. Of course, Ackermann's conditional, which should not be confused with his entailment connective of Ackermann (1956a), has still another "look-and-feel" than the intuitionistic or classical conditional. Ackermann's basic idea is to provide $\to$ with a metalogical reading: "[...] $A \to B$ soll bedeuten, dass bei Zugrundelegung der Formel A die Formel B ableitbar ist", (Ackermann, 1950b, 34). Using Arnold Schmidt's (1960, 268) apt term, we may call this the "inferential interpretation" of the conditional ("Erschließungsdeutung der Implikation"). According to its inferential interpretation, the connective $\to$ is just an object language mirror image of the consequence relation.[11]

Central to Ackermann's version of the inferential interpretation of the conditional is the notion of a formula's semantical level: this is the maximal number of nested occurences of the conditional connective in it. At the bottom, there are the formulas of level 0; e.g., p, $p \wedge q$, $p \vee q$, $\sim p$, etc. (for a language of propositional logic). We express inferential relationships between them by level 1 formulas; for instance, $p \wedge q \to p$ and $p \to p \vee q$. On level 2, we may express the transitivity of the consequence relation[12] by $[p \to q] \wedge [q \to$

---

[11] We leave aside the question whether "consequence" is to be taken in a semantical or syntactical sense. Clearly, Ackermann himself always speaks of derivability. Hence, Ackermann (1965) uses the symbol $\vdash$ rather than $\to$ for the conditional connective.

[12] It might be argued that it would be more consonant with Ackermann's aims to distiguish conditional connectives of different levels. Thus $\to_0$ expresses derivability between object level statements, $\to_1$ derivability between metastatements about object level statements, $\to_2$ derivability between metastatements of the first kind, etc. Ackermann (1950b, 35) admits that we should do this — "streng genommen" — but that he does not want to complicate matters more than necessary. A multi-layered conditional of the kind sketched has been developed by

$r] \to [p \to r]$. A homogeneous formula is a formula built up from formulas belonging to the same semantic level. Each atomic formula is homogeneous; and if A and B are homogeneous formulas of the same semantic level, then $\sim A$, $A \wedge B$, $A \vee B$, $A \to B$ are homogeneous, too. The latter formula, of course, is of higher level than its components. Our previous example of a level 2 formula is homogeneous: its antecedent is a conjunction of two conditionals of level 1; the conjunction itself is thus again of level 1 as is the formula $[p \to r]$ to the right of the main conditional sign. The second level formula $p \to [q \to p]$, however, is inhomogeneous. From the viewpoint of the inferential interpretation, inhomogeneous formulas are hard to interpret because of the difference in semantic level between their components. In our example, the antecedent p may stand for a plain statement on factual matters, whereas $q \to p$ is a metatheoretic statement concerning the consequence relationship. How could the latter be derivable from the first? Thus, though $p \to [q \to p]$ is a valid formula even in mimimal logic, it is rejected in Ackermann's framework.

Some inhomogeneous formulas, however, are undeniably true. From a meta-statement $\top \to A$ saying that A follows from the *verum* $\top$, we conclude without any hesitation that A. Thus we should adopt $[\top \to A] \to A$ though this is inhomogeneous for each formula A. Furthermore, inhomogeneous formulas are easily derived from homogeneous ones using the rule of substitution. Since p is trivially derivable from itself, one will of course adopt the homogeneous $p \to p$. From this, by substitution, we get however the inhomogeneous $[p \to [q \to r]] \to [p \to [q \to r]]$. The latter formula, though completely symmetric in its construction, is nevertheless classified as inhomogeneous by our definition since it contains an inhomogeneous subformula. Thus we cannot completely exclude inhomogeneous formulas. Consequently, Ackermann only requires that "principally" the validity of a conditional formula presupposes that its antecedent and consequent are on the same semantic level. Clearly, the qualification "principally" admits for exceptions. We shall return to these matters in the next section.

Besides the conditional there is another critical connective within Ackermann's system, namely negation. Since Ackermann adopts Behmann's suggestion towards a partial logic, the question arises whether $\sim A$ should amount to the falsehood of A or just to the non-truth of A. The validity of contrapositive arguments depends on this issue (see Ackermann, 1950b, 44). Assume that $\forall x.[P(x) \to Q(x)]$ and $\sim Q(a)$ are both true. If $\sim Q(a)$ is interpreted as stating that $Q(a)$ either is false or lacks a truth value, we conclude the same for $P(a)$. Thus, we should have $\sim P(a)$ in this case. But if $\sim Q(a)$ is the stronger proposition that $Q(a)$ is false, we may only infer from the first premiss that $P(a)$ is not true which does not exclude the possibility of a truth value gap. Thus we are not allowed to conclude $\sim P(a)$ in this case. Ackermann adopts the latter interpretation of negation; thus $\sim A$ means that A is false.

Myhill (1975) and was employed in the formulation of class theory and analysis in Myhill (1984) and Flagg and Myhill (1987).

The negation of compound formulas is reduced to that of their components. Besides partiality, unrestricted comprehension, and the special treatment of the conditional, this is the fourth characteristic of Ackermann's systems (see 1952b, 364). With respect to negated formulas this means that the law of double negation is accepted; conjunctions and disjunctions are treated in accordance to De Morgan's laws. Conditionals are again a problem. Ackermann (1950b) states just a sufficient condition for the falsehood of $A \to B$. The formula $\sim[A \to B]$ is implied by the conjunction of $\top \to A$ and $B \to \bot$.[13] The missing necessary condition for the falsehood of an implication, however, yields a problem in connection with partiality. Admitting partiality means giving equal rights to the notions of truth and falsehood; falsehood does not just mean lack of truth. Of course, then, the explanation of equivalence of formulas as the truth (or derivability) of the biconditional $A \leftrightarrow B$ becomes inadequate. A more adequate explanation would be $[A \leftrightarrow B] \wedge [\sim A \leftrightarrow \sim B]$. However, as Harrop (1954) in his analysis of the original system in Ackermann (1950b) pointed out, the thus defined equivalence relation lacks the usual replacement properties and this is essentially due to the missing necessary condition for the falsity of a conditional. Ackermann (1952b) repairs this flaw[14] by adding $\sim[A \to B] \to [\top \to A] \wedge [B \to \bot]$. This amendment sets the matter straight as is also acknowledged by Harrop (1954, 509).

One dimension of variation in Ackermann's systems of class theory remains still to be mentioned. This regards the question of combining statements lacking a truth value by means of the extensional connectives $\sim$, $\wedge$, and $\vee$ with true and false statements. In the system of Ackermann (1950b), for instance, it is allowed to infer $A \vee B$ from A even when B lacks a truth value. Similarly, one may infer in this system $A \wedge B \to C$ from $A \to C$ even when B is without truth value. This, however, is considered a flaw ("Schönheitsfehler") in Ackermann's last publication on this matter (see Ackermann, 1965, 8). There a special inference rule for propositional logic is adopted which ensures that only such theorems are derivable which are built up from sentences known to have a truth value. — In the following, we shall not deal with all systems proposed by Ackermann. Instead we concentrate on the system resulting from that developed in Ackermann (1950b) when the revision due to Harrop (1954) is added to it; this is the "core" of the system in Ackermann (1952b). We shall label this system by the letter **A**. **A** is formulated within a first-order language comprising (besides the paratheses) (1) the propositional connectives $\top$, $\bot$, $\sim$, $\wedge$, $\vee$, and $\to$, (2) the quantifiers $\forall$ and $\exists$, (3) the special signs $=$, $\{\,|\,\}$, and $\in$, and (4) individual variables.[15] As is common in the Hilbert school, Ackermann typographically distinguishes bound variables x, y, z, x_1, ... from

---

[13] Ackermann uses Γ instead of $\top$ and $\sim\Gamma$ instead of $\bot$.
[14] Thus reacting to Harrop Harrop (1954) — as admitted by Ackermann (1958, 6, fn 1).
[15] Ackermann (1950b) defines the verum $\top$ by the formula which states that the identity relation is identical to itself: $= = =$. Furthermore, he writes $\hat{x}_1 \hat{x}_2 \ldots \hat{x}_n A(x_1, x_2, \ldots, x_n)$ — in later articles $\lambda x_1 x_2 \ldots x_n A(x_1, x_2, \ldots, x_n)$

free parameters $a, b, c, a_1, \ldots$. The deductive apparatus of **A** is displayed in Tab. 1. The quantificatonal skemes A20–A24 are object to the usual resrictions, namely that x may not occur in A (in A21 and A22), respectively in B (in A23).

4 Σ-Logic

In the different expositions of his systems of class theory, Ackermann provides Hilbert type axiomatic systems for them.[16] In order to give a more motivated account, we shall deviate from this practice giving here a special sequent formulation for **A**. This sequent formulation is inspired by, though different from Ackermann's Σ-formulations of classical and entailment logic given in Ackermann (1956a), cf. also Anderson et al. (1992, ch. VIII). Nevertheless, I will call this variety of the sequent approach "Σ-logic". Σ-logic makes the motivation backing the propositional part of **A** very transparent and makes it easy to relate it to other logics.

4.1 Homogeneous Σ-Logic

According to Ackermann's inferential interpretation of the conditional, $A \to B$ means that B is derivable from A. In Σ-logic, the latter relation is encoded by the sequent $A \Rightarrow B$. Σ-logic is a framework to handle such sequents. There are three kinds of principles: (1) Principles stating the basic assumptions about the consequence relation. These may be either "purely structural" principles such as the cut rule or principles relating to the conditional. (2) Principles stating the basic consequence relationships involving the connectives $\bot$, $\top$, $\wedge$, and $\vee$. (3) Principles for mirroring sequents by conditionals sentences. I will call these three sets of principles the "consequence module", the "connective module", and the "translation module", respectively.

Sequents in Σ-logic are conceived of as linear versions of rules of inference. If we have conjunction at our disposal, we can "normalize" all rules to such having at most one premiss.[17] Of course, in order to justify such a reduction, we make use of the rules $A \wedge B \Rightarrow A$ and $A \wedge B \Rightarrow B$ as well as of $A, B \Rightarrow A \wedge B$. But then, all we need are rules with maximally two premisses — thus sequents with at most two formulas in their antecedent. The sequents allowed by Ackermann (1956a) in his Σ-systems are of just this kind;

— instead of $\{x_1 x_2 \ldots x_n | A(x_1, x_2, \ldots, x_n)\}$ and $\{a\}(b_1 b_2 \ldots b_n)$ instead of $(b_1 b_2 \ldots b_n) \in a$.

[16] For the sake of metamathemical investigation, Ackermann (1952b, §3) uses "disjunction sequences" for formulating class theory. These are sequences of formulas which are to be read disjunctively, thus classical Gentzen sequents with empty antecedents. The system in Ackermann (1965, §2) is a Hilbert system enriched by a natural deduction rule for the introduction of conditionals.

[17] Axioms are conceived as the consequents of premissless rules.

Ackermann's System A of Class Theory

I. Propositional Logic

A1	$A \to A$	A2	$[A \to B] \land [B \to C] \to [A \to C]$
A3a	$A \land B \to A$	A3b	$A \land B \to B$
A4	$[A \to B] \land [A \to C] \to [A \to B \land C]$		
A5a	$A \to A \lor B$	A5b	$B \to A \lor B$
A6	$[A \to C] \land [B \to C] \to [A \lor B \to C]$		
A7	$A \land [B \lor C] \to B \lor [A \land C]$		
A8	$A \to \top$	A9	$[\top \to A] \to A$
A10	$[\top \to A] \to [B \to A \land B]$	A11	$[\top \to [A \lor B]] \to [\top \to A] \lor [\top \to B]$
A12	$\bot \to A$	A13	$A \land {\sim}A \to \bot$
A14a	$\bot \to {\sim}\top$	A14b	${\sim}\top \to \bot$
A15a	$\top \to {\sim}\bot$	A15b	${\sim}\bot \to \top$
A16a	$A \to {\sim}{\sim}A$	A16b	${\sim}{\sim}A \to A$
A17a	${\sim}A \lor {\sim}B \to {\sim}[A \land B]$	A17b	${\sim}[A \land B] \to {\sim}A \lor {\sim}B$
A18a	${\sim}A \land {\sim}B \to {\sim}[A \lor B]$	A18b	${\sim}[A \lor B] \to {\sim}A \land {\sim}B$
A19a	$[\top \to A] \land [B \to \bot] \to {\sim}[A \to B]$		
A19b	${\sim}[A \to B] \to [\top \to A] \land [B \to \bot]$		

II. Quantification Theory

A20	$\forall x.A(x) \to A(a)$	A21	$\forall x.[A \to B(x)] \to [A \to \forall x.B(x)]$
A22	$\forall x.[A \lor B(x)] \to A \lor \forall x.B(x)$	A23	$\forall x.[A(x) \to B] \to [\exists x.A(x) \to B]$
A24	$A(a) \to \exists x.A(x)$	A25	$A \land \exists x.B(x) \to \exists x.[A \land B(x)]$
A26	$[\top \to \exists x.B(x)] \to \exists x.[\top \to B(x)]$		
A27a	$\exists x.{\sim}A(x) \to {\sim}\forall x.A(x)$	A27b	${\sim}\forall x.A(x) \to \exists x.{\sim}A(x)$
A28a	$\forall x.{\sim}A(x) \to {\sim}\exists x.A(x)$	A28b	${\sim}\exists x.A(x) \to \forall x.{\sim}A(x)$

III. Identity

A29	$a = a$	A30	$a = b \land A(a) \to A(b)$
A31	$A(a) \land {\sim}A(b) \to a \neq b$		

IV. Class Theory

A32a $A(a_1, a_2, \ldots, a_n) \to (a_1, a_2, \ldots, a_n) \in \{(x_1, x_2, \ldots, x_n) | A(x_1, x_2, \ldots, x_n)\}$
A32b $(a_1, a_2, \ldots, a_n) \in \{(x_1, x_2, \ldots, x_n) | A(x_1, x_2, \ldots, x_n)\} \to A(a_1, a_2, \ldots, a_n)$
A33a ${\sim}A(a_1, a_2, \ldots, a_n) \to (a_1, a_2, \ldots, a_n) \notin \{(x_1, x_2, \ldots, x_n) | A(x_1, x_2, \ldots, x_n)\}$
A33b $(a_1, a_2, \ldots, a_n) \notin \{(x_1, x_2, \ldots, x_n) | A(x_1, x_2, \ldots, x_n)\} \to {\sim}A(a_1, a_2, \ldots, a_n)$

Rules of Inference

Modus ponens, the rule allowing to infer $\top \to A$ from A, change of bound variables, universal generalization and substitution of terms for free variables.

Table 1. The Hilbert system **HA** of Ackermann's class theory.

furthermore, Ackermann admits only sequents having at most one formula in their succedents. Allowing for sequents with an empty succedent already corresponds to a slight generalization of the notion of an inference rules. One thereby admits also rules stating the inconsistency of some formulas. Here we go one step further than Ackermann and admit also multiple conclusion rules. Treating more than two conclusion by means of disjunction in an analogous way as premisses by means of conjunction, we may restrict ourselves to sequents with at most two formulas in their succedent. Thus we arrive at the notion of a Σ-sequent: this is a sequent $\Gamma \Rightarrow \Delta$ with $|\Gamma|, |\Delta| \leq 2$.

Σ-sequents are linearizations of inference rules, and these linearization are then converted further by the translation module into conditional formulas. Two translation procedures are well known from the literature. Let $\Gamma^\wedge$ (resp. $\Delta^\vee$) be $\top$ (resp. $\bot$) if Γ (resp. Δ) is empty, A if Γ (resp. Δ) consists of just this formula, and $A \wedge B$ (resp. $A \vee B$) if Γ (resp. Δ) comprises the two formulas A and B. Then the first translation of $\Gamma \Rightarrow \Delta$ is $\Gamma^\wedge \Rightarrow \Delta^\vee$. This was already given by Gentzen (1934/35, 180) as an explanation of his sequent notation. In Σ-logic, as we will see soon, this translation procedure is directly implemented by means of special rules. For the second translation, we declare $\tilde{\Gamma}(\Gamma, A)$ to be $\top \to A$ in case that Γ is empty, $B \to A$ if Γ is just the formula B, and $C \to [B \to A]$ if Γ is B, C. The second translation of $\Gamma \Rightarrow \Delta$, then, is $\tilde{\Gamma}(\Gamma, \Delta^\vee)$. The second procedure relies on an iterated application of Gentzen's rule (FES) for introducing a conditional into a succedent.

$$(\text{FES}) \quad \frac{A, \Gamma \Rightarrow \Theta, B}{\Gamma \Rightarrow \Theta, A \to B} \qquad (3)$$

This rule allows to empty the antecedent by making the members of the succedent conditionally dependent on the antecedent formulas. Since (FES) essentially is a sequent implementation of the deduction theorem, this rule is, as may be expected from our discussion of Curry's paradox in section 3.1 above, not generally available within Σ-logic. Furthermore, Σ-logic was invented by Ackermann for representing his logic of entailment; and there, too, the general deduction theorem fails as it does in **A** (see Anderson and Belnap, 1975, 256–261). However, though the deduction theorem generally fails within some logics representable within the Σ-framework, there are sometimes special circumstances where it is valid. In order to have control over these matters, Ackermann (1956a, 117f) introduced his "star device". If in the antecedent of a Σ-sequent a formula is attached a star, this means that (FES) may be applied to the neighbouring fomula. Thus, for example[18], $A, B^* \Rightarrow C$ gives rise to $B \Rightarrow A \to C$.

Table 2 (page 37 below) displays the principles of the system **ΣHP** of homogeneous propositional Σ-logic.[19] The consequence module comprises the two parts of the cut rule, which is the only "structural" principle of **ΣHP**.

---

[18] The example given cannot arise in Ackermann's Σ-formulation of his logic of entailment. There the starred formula is always to the left of the unstarred one.

[19] All the rules of Table 2 are of course restricted by the requirement that they lead from Σ-sequents as premisses again to such a sequent as a conclusion. Thus, for instance, Γ in (⊤Intr) is restricted by the condition $|\Gamma| \leq 1$

The left subrule of of (Cut) "cancels" the only premiss of a singulary rule; it does not, however, allow for "cutting out" a single proved item from a sequent with two antecedent formulas. There is no general way to procede from $\Rightarrow A$ and $A, B \Rightarrow \Gamma$ to $B \Rightarrow \Gamma$. For binary sequents, however, we have the following derived version of the cut rule.

Observation 1 *In* **ΣHP** *there are the derived rules*

$$\frac{\Rightarrow A \quad \Rightarrow B \quad A, B \Rightarrow \Gamma}{\Rightarrow \Gamma} \quad \text{and} \quad \frac{\Gamma \Rightarrow A, B \quad A \Rightarrow \quad B \Rightarrow}{\Gamma \Rightarrow}. \qquad \Box$$

Proof. The left hand subrule is treated as follows:
(1) $\Rightarrow A$ hypothesis
(2) $\Rightarrow B$ hypothesis
(3) $\Rightarrow A \wedge B$ 1, 2, (Intr$\wedge$)
(4) $A, B \Rightarrow \Gamma$ hypothesis
(5) $A \wedge B \Rightarrow \Gamma$ 4, ($\wedge\downarrow$)
(6) $\Rightarrow \Gamma$ 3, 5 (Cut)

The right hand rule is derived analogously with the help of disjunction instead of conjunction. ⊞

Leaving the explanation of the consequence module's other rules for later, we continue instead with the rules of the other modules. The connective module comprises the usual Gentzen rules for conjunction and disjunction. The special constants *verum* ($\top$) and *falsum* ($\bot$) may be interpreted respectively as the conjunction of all logical truths and the disjunction of all logical falsehoods, thus as the relevant logician's *t* and *f* (see Anderson and Belnap, 1975, 342). The translation module comprises the so-called Ketonen rules for conjunction and disjunction (see Ketonen, 1944, 14) together with their converses. The double line rules ($\wedge\updownarrow$) and ($\vee\updownarrow$) may be applied from top to bottom (these are the Ketonen cases ($\wedge\downarrow$) and ($\vee\downarrow$)) and bottom up (their converses ($\wedge\uparrow$) and ($\vee\uparrow$)). Since the rules ($\top$Intr) and (Intr$\bot$) may be used to fill up empty antecedents and succedents, the translation rules allow the conversion of any sequence $\Gamma \Rightarrow \Delta$ into $\Gamma^\wedge \Rightarrow \Delta^\vee$. A single step, then, remains to convert this further into $\Rightarrow \Gamma^\wedge \to \Delta^\vee$. This step is effected by the complicated rule ($\to\star$), which in the framework of **ΣHP** replaces Gentzens rule (FES). Postponing the discussion of the translation module's last rule (Elim$\to$), we are going to explain the rule ($\to\star$) first.

Note first that all the rules discussed until now are homogeneous in that sense that they procede from sequents relating homogeneous formulas of the same semantic level as premisses to another sequent of this type as a conclusion. More precisely, every instance of such a rule can be derived *via* substitution from a "base instance" using propositional variable instead of schematic letters. All formulas occuring in a sequent of a base instance are homogeneous and of the same semantic level. Thus, for instance, the rule (Intr$\wedge$) can be encoded as in (4).

Table 2. Basic propositional Σ-Logic

The System ΣHP of Homogeneous Propositional Σ-Logic

The consequence module

(Cut) $\dfrac{\Rightarrow A \quad A \Rightarrow \Gamma}{\Rightarrow \Gamma}$ 　　　　(Cut) $\dfrac{\Gamma \Rightarrow A \quad A \Rightarrow}{\Gamma \Rightarrow}$

($\rightarrow$Ref) $\Rightarrow A \rightarrow A$

($\rightarrow\star$) $\dfrac{\Gamma^{(*)} \Rightarrow \Delta^{(*)}}{(A\rightarrow)\Gamma \Rightarrow (A\rightarrow)\Delta}$ 　　　($\star\rightarrow$) $\dfrac{\Gamma \Rightarrow \Delta}{\Delta \rightarrow A \Rightarrow \Gamma \rightarrow A}$

Conditions: 　　　　　　　　　　　　**Conditions:**

1. $|\Delta| = 1$.　　　　　　　　　　　　1. $|\Gamma| = 1$.
2. $|\Gamma| \geq 1$.　　　　　　　　　　　2. $|\Delta| \geq 1$.
3. Each unstarred formula of Γ and Δ is attached the prefix $A \rightarrow$.
4. No starred formula receives this prefix.

The connective module

($\top$Intr) $\dfrac{\Gamma \Rightarrow \Delta}{\top, \Gamma \Rightarrow \Delta}$ 　　　　　　(Intr$\top$) $\Rightarrow \top$

($\bot$Intr) $\bot \Rightarrow$ 　　　　　　　　　(Intr$\bot$) $\dfrac{\Gamma \Rightarrow \Delta}{\Gamma \Rightarrow \Delta, \bot}$

(Intr$\wedge$) $\dfrac{A, \Gamma \Rightarrow \Delta}{A \wedge B, \Gamma \Rightarrow \Delta}$ 　　　(Intr$\wedge$) $\dfrac{\Gamma \Rightarrow \Delta, A \quad \Gamma \Rightarrow \Delta, B}{\Gamma \Rightarrow \Delta, A \wedge B}$

(Intr$\wedge$) $\dfrac{B, \Gamma \Rightarrow \Delta}{A \wedge B, \Gamma \Rightarrow \Delta}$

($\vee$Intr) $\dfrac{A, \Gamma \Rightarrow \Delta \quad B, \Gamma \Rightarrow \Delta}{A \vee B, \Gamma \Rightarrow \Delta}$ 　(Intr$\vee$) $\dfrac{\Gamma \Rightarrow \Delta, A}{\Gamma \Rightarrow \Delta, A \vee B}$

　　　　　　　　　　　　　　　　　　(Intr$\vee$) $\dfrac{\Gamma \Rightarrow \Delta, B}{\Gamma \Rightarrow \Delta, A \vee B}$

The translation module

($\wedge\updownarrow$) $\dfrac{A, B \Rightarrow \Delta}{A \wedge B \Rightarrow \Delta}$ 　　　　　($\vee\updownarrow$) $\dfrac{\Gamma \Rightarrow A, B}{\Gamma \Rightarrow A \vee B}$

(Elim$\rightarrow$) $\dfrac{\Gamma \Rightarrow \Delta, A \rightarrow B}{A, \Gamma^* \Rightarrow \Delta^*, B}$

$$\dfrac{(p)\,(q) \Rightarrow (r)\,s \quad (p)\,(q) \Rightarrow (r)\,t}{(p)\,(q) \Rightarrow (r)\,s \wedge t} \tag{4}$$

(Components in brackets are facultative.) Gentzen's rule (FES), displayed above in (3), being inhomogeneous does not possess such an encoding and is to be rejected.[20]

There are two basic ideas behind (FES). The first is to derive a new sequent from a given one $A, \Gamma \Rightarrow \Theta, B$ by recognizing an additional condition C on which B dependends as well as some of the assumptions in A, Γ. Given that the order of the assumptions does not matter, we may assume that it is A which also depends on C, thus $C \to A, \Gamma \Rightarrow \Theta, C \to B$. The second idea, then, is to identify C and A. Presupposing the availability of an unrestricted cut rule, we may cancel the trivial $A \to A$ in the antecedent and get the conclusion $\Gamma \Rightarrow \Theta, A \to B$ of (FES). The first idea can easily be adjusted to Ackermann's homogenity principle. We just attach the prefix $C \to$ to every formula of the premiss sequent in order to conserve homogeneity. Let us denote by $A \to \Gamma$ (resp.: $\Gamma \to A$) the sequence of formulas which results from Γ by attaching the prefix (resp.: suffix) $A \to$ (resp.: $\to A$) to each of its members. Of course, the new sequence will be empty if the old sequence Γ is. Thus the rule which is to replace Gentzen's (FES) should precede from a sequent $\Gamma \Rightarrow \Delta$ to the conclusion $A \to \Gamma \Rightarrow A \to \Delta$.

For the thus revised rule to be valid, we should require that $|\Delta| \geq 1$. Even then, however, two caveats remain. We easily derive $B \vee C \Rightarrow B, C$ in **ΣHP**. Making antecedent and succedent dependent on A yields $A \to B \vee C \Rightarrow A \to B, A \to C$ which is hardly compatible with our inferential reading of the arrow. Being able to infer from a novel's displaying stylistic feature A that it originates either from writer B or from writer C is completely compatible with neither being able to infer writer B from trait A nor being able to conclude C's authorship from A; cp. Schmidt (1960, 268) for this kind of counterexamples. Thus we have to postulate that $|\Delta| = 1$. The second objection concerns matters of relevance. Admitting an empty antecedent Γ would allow making the succedent formula depending on an irrelevant condition A. We exclude this by requiring $|\Gamma| \geq 1$. Employing, as in (4), basic instances instead of schemata, we get (5).

$$\frac{p(q) \Rightarrow r}{s \to p(s \to q) \Rightarrow s \to r} \qquad \frac{p \Rightarrow q(r)}{q \to s(r \to s) \Rightarrow p \to s} \tag{5}$$

As one easily recognizes, there are no quarrels about homogenety with this.

However, we are not yet finished with the revision of Gentzen's (FES). This rule comes along with a counterpart (FEA) dealing with conditionals in the antecedent; (Gentzen, 1934/35, 193).

$$\text{FEA} \ \frac{\Gamma \Rightarrow \Theta, A \quad B, \Delta \Rightarrow \Lambda}{A \to B, \Gamma, \Delta \Rightarrow \Theta, \Lambda} \qquad \text{FES}_\text{i} \ \frac{\Gamma \Rightarrow \Theta, A \to B}{A, \Gamma \Rightarrow \Theta, B} \tag{6}$$

Assuming unrestricted cut, (FEA) is equivalent to the inverse (FES$_\text{i}$) of the rule (FES). We concentrate on the latter rule, since for $|\Gamma, \Theta| = 0$, it codifies

---

[20] The special case of $|\Gamma, \Theta| = 0$ is unobjectable and derivable in **ΣHP**, see below Observation 2(c).

one half of the inferential interpretation of the conditional. Like (FES), the rule (FES$_i$) destroys homogenity. However, there is nothing to complain about the inhomgeneous sequents arising from applications of this rule if we interpret the conclusion sequent as a mere notational variant of the premiss sequent. Thus we do not translate the conclusion of this rule by the formula $(A, \Gamma)^\wedge \to (\Gamma, B)^\vee$ but use instead the same formula translation as for the premiss. In order to avoid ambiguity of notation, Ackermann's star device mentioned above is used to indicate this special interpretation. We mark the formulas in Γ (and Δ) by an asterisk.[21] Thus the conclusion sequent would be $A, \Gamma^* \Rightarrow \Theta^*, B$ where the asterisk means that each formula in Γ and Θ receives this mark. The thus modified rule (FES$_i$) is called (Elim$\to$) here. The applications of (Elim$\to$) are the substitution instances of (7).

$$\frac{(p_1 \to p_2) \Rightarrow (q_1 \to q_2) \; r \to s}{r \;([p_1 \to p_2]^*) \Rightarrow ([q_1 \to q_2]^*) \; s} \quad (7)$$

Again, there are no quarrels about homogenity with this since the stars in the conclusion sequent force an intepretation of r as a condition of s. Thus the rule is purely "analytic". The same holds true for the rules $(\wedge\updownarrow)$ and $(\vee\updownarrow)$, and we take this kind of analyticity as the hallmark of translational rules. Therefore (Elim$\to$) belongs to the translation module.

After having introduced the star as a new notational device, we have to reconsider our old rules. With one exception, we generally exclude stars from these rules. The single exception concerns the rule $(\to\star)$. Here we easily recognize that we can restore formal homogeneity in a conclusion sequent by prefixing the formulas deriving from r and s in (7) by prefixing them by a suitable condition. This does not only restore formal homogenity, this transition is also logically sound when we cancel the stars in the conclusion. Thus we want to have (8) as a further subrule of $(\to\star)$.

$$\frac{p \;([q_1 \to q_2]^*) \Rightarrow ([r_1 \to r_2]^*) \; s}{t \to p \;(q_1 \to q_2) \Rightarrow (r_1 \to r_2) \; t \to s} \quad (8)$$

There is yet the rule $(\star\to)$ to explain. The essential idea of $(\to\star)$ was to make a sequent's antecendent formulas dependent on an additional hypothesis A. Since a conditional possesses a consequent besides its hypothesis, we may ask what dually happens when the succedent formulas of a sequent $\Gamma \Rightarrow \Delta$ each give rise to a common consequent A. In this case, the antecedent Γ collectively has the very same consequence A. Thus we have $\Delta \to A \Rightarrow \Gamma^\wedge \to A$. Since we do not want to employ the notion of conjunction in a rule dealing with the conditional, we have no chance than to require $|\Gamma| = 1$ to get the rule straight. An empty Δ is excluded because of relevance considereations.

In Tab. 2, we use Greek capitals instead of indicating contexts by means of propositional variables and special formulas. A plain capital Greek letter

---

[21] Since Ackermann does not consider succedents with more than one formula, he does not need to employ the star device in the succedent.

denotes a sequence of maximally two unstarred formulas. If the Greek letter is starred, each of its constitutent formulas is thus marked. Brackets around the star mean that the formulas of this sequence may be either starred or unstarred. By "$(A \to)\Gamma$" we denote any sequence deriving from Γ by prefixing no, some, or all of its formulas by $A \to$. In using Greek letters and formula schemes instead of basic rule instances as in (5), (7), and (8), we are at the same time generalizing our rules. The latter rule formulations are only used to make clear that the rules involved are not problematic with respect to homogenity if one always applies rules in "reasonable way", i.e., by taking care of matters of homogenity. But we do not obligatorily require such a careful use of rules in Σ**HP**. We may, for instance, start a derivation with an instance $\Rightarrow p \to p$ of ($\to$Ref). Applying ($\vee$Intr) to this, we get $\Rightarrow q \vee [p \to p]$. Though this is unreasonable given Ackermann's homogenity requirement, it is hardly to avoid without adding cumbersome restrictions to the formulations of (Intr$\vee$). The mentioned sequent offends Ackermann's homogenity principle, but it is harmless. There is thus no much reason in being so explicit about the side formula's structure as in (7) and (8). Therefore we accept for Σ**HP** the more general rule formulations given in Table 2.

4.2 Comparing A with ΣHP

Σ**HP** characterizes the domain of propositions as a lattice ordered by the relation $\Rightarrow$. Besides the lattice operations $\wedge$ and $\vee$, there is a special operation mapping instances $A \Rightarrow B$ of the lattice order onto lattice elements $A \to B$. In Observation 2(a) and (d) below, we observe that $\Rightarrow$ is reflexive and transitive (as it should be as a lattice order).

Observation 2 *In Σ**HP** the following sequents and rules are derivable:*

(a) $A \Rightarrow A$,
(b) $A \to B, B \to C \Rightarrow A \to C$,
(c) $\dfrac{A \Rightarrow B}{\Rightarrow A \to B}$,
(d) $\dfrac{A \Rightarrow B \quad B \Rightarrow C}{A \Rightarrow C}$.

Proof. (a) This follows from ($\to$Ref) by an application of (Elim$\to$).

(b) (1) $B \to C \Rightarrow B \to C$ (a)
    (2) $B, [B \to C]^* \Rightarrow C$ 1, (Elim$\to$)
    (3) $A \to B, B \to C \Rightarrow A \to C$ 2, ($\to\star$)

(c) (1) $A \Rightarrow B$ hyp.
    (2) $A \to A \Rightarrow A \to B$ 1, ($\to\star$)
    (3) $\Rightarrow A \to B$ ($\to$Ref), (2), (Cut)

(d) (1) $A \Rightarrow B$ Hyp.
    (2) $\Rightarrow A \to B$ (1), (c)
    (3) $B \Rightarrow C$ Hyp.
    (4) $\Rightarrow B \to C$ (3), (c)
    (5) $\Rightarrow [A \to B] \wedge [B \to C]$ (2), (4), ($\wedge$Intr)
    (6) $[A \to B] \wedge [B \to C] \Rightarrow A \to C$ from (b) by ($\wedge\downarrow$)
    (7) $\Rightarrow A \to C$ (5), (6), (Cut)
    (8) $A \Rightarrow C$ (7), (Elim$\to$)

The proof of (d) nicely illustrates the difference in "look and feel" between the classical Gentzen framework and Ackermann's Σ-logic. While in the former the connectives derive part of their strength from the structural rules, in Σ-logic purely structural principles, as the cut principle (d), depend on the properties of the connectives. — It is not so difficult to show by means of the rules of the connective module that conjunction and disjunction have the usual lattice properties. Thus the conjunction (disjunction) of two propositions is a lower (upper) bound of them, i.e., we may derive $A \wedge B \Rightarrow A$, $A \Rightarrow A \vee B$, etc.). Applying the derived rule from Obs2(c) to these sequents, we get Axioms A3a, A3b, A5a, and A5b of Ackermann's system **A**; cf. Table 1. Axioms A2 and A4 are treated in the following observation.

Observation 3 *In* Σ**HP** *we have:*

(a) $\Rightarrow [A \to B] \wedge [A \to C] \to [A \to B \wedge C]$,
(b) $\Rightarrow [A \to C] \wedge [B \to C] \to [A \vee B \to C]$.

Proof. (a) (1) $B \wedge C \to B \wedge C$ Obs 2(a)
       (2) $B, C \Rightarrow B \wedge C$ (1), ($\wedge\uparrow$)
       (3) $A \to B, A \to C \Rightarrow A \to B \wedge C$ (2), ($\to\star$)
       (4) $[A \to B] \wedge [A \to C] \Rightarrow A \to B \wedge C$ (3), ($\wedge\downarrow$)
       (5) $\Rightarrow [A \to B] \wedge [A \to C] \to [A \to B \wedge C]$ 4, Obs2(c)

(b) is treated dually to (a).

In relevance logic, distributivity (of conjunction over disjunction) comes along as a somewhat unexpected addition; cp. Došen (1993, 11) on this matter. There is, however, nothing anormalous with the derivation of Ackermann's distributivity axiom A6 in Σ**HP**.

Observation 4 *In* Σ**HP** *we have*

(a) *the derived rule* $\dfrac{A, B \Rightarrow \Gamma}{B, A \Rightarrow \Gamma}$ *and*

(b) *the derivable sequent* $\Rightarrow A \wedge [B \vee C] \to B \vee [A \wedge C]$.

Proof. (a) is easily shown by first deriving $B \wedge A \Rightarrow A \wedge B$ and then using ($\wedge\updownarrow$) and (Cut).

With respect to (b), we proceed as follows:

(1) $B \Rightarrow B \vee [A \wedge C]$	Obs2(a), (Intr$\vee$)
(2) $B, A \Rightarrow B \vee [A \wedge C]$	1, ($\wedge$Intr), ($\wedge\uparrow$)
(3) $C \wedge A \Rightarrow A$	Obs2(a), ($\wedge$Intr)
(4) $C \wedge A \Rightarrow C$	Obs2(a), ($\wedge$Intr)
(5) $C, A \Rightarrow A \wedge C$	3, 4, (Intr$\wedge$), ($\wedge\uparrow$)
(6) $C, A \Rightarrow B \vee [A \wedge C]$	5, (Intr$\vee$)
(9) $[B \vee C], A \Rightarrow B \vee [A \wedge C]$	2, 6, ($\vee$Intr)
(10) $\Rightarrow A \wedge [B \vee C] \to B \vee [A \wedge C]$	9, (a), ($\wedge\downarrow$), Obs2(c)

⊞

Let us call a formula A derivable in **ΣHP** if the sequent $\Rightarrow A$ is derivable in that system. Then we observe:

Observation 5 *The set of formulas derivable in* **ΣHP** *is closed under*

(a) modus ponens and
(b) the rule to infer $\top \to A$ *from* A.   ⊟

Proof. (a) From the hypothesis $\Rightarrow A \to B$ we derive by (Elim$\to$) the sequent $A \Rightarrow B$ from which we infer $\Rightarrow B$ by (Cut) and the hypothesis $\Rightarrow A$.

(b) We infer the sequent $\top \Rightarrow A$ from the hypothesis $\Rightarrow A$ by ($\top$Intr). Then, applying ($\to\star$) we get $\top \to \top \Rightarrow \top \to A$. From this the conclusion follows by ($\to$Ref) and (Cut).   ⊞

Thus we see that **ΣHP** is a fragment of the positive propositional part of **A**. Conversely, it is not too hard an exercise to show that the sequent rules of **ΣHP** are converted into derived rules of **A** when we translate sequents $\Gamma \Rightarrow \Delta$ by formulas $\Gamma^\wedge \to \Delta^\vee$. We omit details for reasons of space. If we look on the principles of **A** which are lacking from **ΣHP**, then one obvious thing to note is that Ackermann's system characterizes $\top$ and $\bot$ as the maximal and minimal element of the lattice of propositions, cp. Axioms A8 and A12 in Table 1. In **ΣHP**, $\top$ is, as Obs5(b) tells us, minimal in the set of all logical truths. This is consonant with equating it with the relevant logician's t (the conjunction of all logical truths), but it does not yet provide $\top$ with the status of a maximal element (the relevant logician's T, *i.e.*, the disjunction of all propositions, cp. Anderson and Belnap (1975, 342)). Another issue is that **ΣHP** — being even more "homogeneous" than Ackermann's **A** — does not allow for the derivation of axiom A8, which enables one to "step down" the hierarchy of semantic levels in specific cases. Furthermore, both the "modified exportation principle" A10 and the distribution principle A11 are missing from **ΣHP**.

4.3 From ΣHP to A (and further)

Up to now, the only rule allowing to get rid of stars is ($\to\star$). The inferential interpretation of the conditional, however, suggests another rule of this kind. Suppose that $\Gamma \Rightarrow \Delta, A \to B$ and that $\Rightarrow A$. Given the conditions Γ, then, it should be possible either to derive an item of Δ or B. By first "transporting A

to the antecedent" by (Elim→), we can license this inference by generalizing our restricted cut rule in the following way.

$$(\text{SCut}) \frac{\Rightarrow A \quad A, \Gamma^* \Rightarrow \Delta^*}{\Gamma \Rightarrow \Delta} \qquad (\text{SCut}) \frac{\Gamma \Rightarrow \Delta, A \quad A \Rightarrow}{\Gamma \Rightarrow \Delta} \qquad (9)$$

Actually, this version of cut is already used by Ackermann (1956a, 117), cp. his rule II'b. However, this generalization of $\Sigma\mathbf{HP}$, which we shall call "G_1", deviates from the strict principle of homogenity by rendering the "downwards stepping axiom" A9 derivable.

Observation 6 *Under G_1 we have:*

(a) $\top \to A \Rightarrow A$ *and*
(b) $\Rightarrow [\top \to A] \to A.$ ⌑

Proof. Since (b) follows from (a) by Obs2(c), it suffices to show (a). This sequent derives from (Intr⊤) and $\top, [\top \to A]^* \Rightarrow A$ by (SCut). $\top, [\top \to A]^* \Rightarrow A$, in turn, is inferred by (Elim→) from an instance of (→Ref).   ⊞

Our next concern is $\top$ and $\bot$. In order to turn them into the top and bottom element of our lattice of propositions, we liberalize the rules (→⋆) and (⋆→) by cancelling in each of these rules the respective second restriction. This generalization is called "G_2".

Observation 7 *Modifying (→⋆) and (⋆→) by G_2 renders the following sequents derivable:*

(a) $A \Rightarrow \top,$
(b) $\Rightarrow A \to \top,$
(c) $\bot \Rightarrow A,$
(d) $\Rightarrow \bot \to A,$
(e) $A \Rightarrow A \wedge \top,$
(f) $A \vee \bot \Rightarrow A.$ ⌑

Proof. We derive (a) from (b) and (c) from (d) by (Elim→). The sequents (b) and (c) imediately follow from respectively (Intr⊤) and (⊥Intr) by the liberalized rules. From (a) and $A \Rightarrow A$ (Obs2(a)) we derive (e) by (Intr∧); and (c) yields (f) analogously by applying (∨Intr).   ⊞

Obs7(b) and (d) correspond respectively to axioms A8 and A12 from **A**. From the viewpoint of relvance logic, it is legitimate to conclude $\top \to A$ from a logically true formula A if $\top$ is interpreted as the conjunction of all logical thruths. Similarly, $A \to \top$ is true if $\top$ is interpreted as the disjunction of all propositions. These are, however, two different interpretation of the sign "⊤" which, for the relevant logician, exclude each other. By adopting G_3, we have erased (or confused, as the relevance logician would put it) this distinction. That we have definitely left the ground of "relevantism" by accepting G_2 becomes apparent by the following observation which states that with G_2 unrestrictive dilution, thus the addition of irrelevant material, becomes available.

Observation 8 *Under* G_2, *we have:*

(a) $\dfrac{\Gamma \Rightarrow \Delta}{A, \Gamma \Rightarrow \Delta}$,

(b) $\dfrac{\Gamma \Rightarrow \Delta}{\Gamma \Rightarrow \Delta, A}$. $\square$

Proof. We just consider (a), the proof (b) being analogous. If Γ consists of a single formula, then we just use ($\wedge$Intr) and ($\wedge\uparrow$). Thus the only case which remains to be considered is that of an empty Γ. Our premiss sequent, then, is $\Rightarrow \Delta$. We procede as follows:

(1) $\Rightarrow \Delta$ premiss
(2) $\top \Rightarrow \Delta$ 1, ($\top$Intr)
(3) $A \wedge \top \Rightarrow \Delta$ 2, ($\wedge$Intr)
(4) $A \wedge \top \Rightarrow \Delta^\vee$ 3, (Intr$\bot$), possibly (if $|\Delta| = 2$) ($\vee\downarrow$)
(5) $A \Rightarrow \Delta^\vee$ 4, Obs7(e), Obs2(d)
(6) $A \Rightarrow \Delta$ 5, possibly ($\vee\uparrow$), or (($\bot$Intr) plus (Cut))

$\boxplus$

Looking quite innocently at first glance, G_2 makes available full cut.

Observation 9 *Under* G_2, *the following cut rules becomes derivable:*

(a) (PCut) $\dfrac{\Rightarrow A \quad A, \Gamma \Rightarrow \Delta}{\Gamma \Rightarrow \Delta}$ (PCut) $\dfrac{\Gamma \Rightarrow \Delta, A \quad A \Rightarrow}{\Gamma \Rightarrow \Delta}$.

(b) (GCut) $\dfrac{\Gamma \Rightarrow \Delta, A \quad A, \Theta \Rightarrow \Lambda}{\Gamma, \Theta \Rightarrow \Delta, \Lambda}$. $\square$

Proof. (a) The proof of the right hand subrule of (PCut) is dual to that of the left hand part. Thus we confine ourselves to the latter. If, furthermore, Γ is empty, that subrule boils down to the corresponding subrule of (Cut). Thus we assume Γ to consist of the formula B. We proceed then as follows.

(1) $A, B \Rightarrow \Delta$ Hyp.
(2) $A \wedge B \Rightarrow \Delta^\vee$ 1, ($\wedge\downarrow$), possibly (Intr$\bot$) or ($\vee\downarrow$)
(3) $B \Rightarrow A$ Hyp, Obs8(a)
(4) $B \Rightarrow A \wedge B$ (3), Obs2(a), ($\wedge$Intr)
(5) $B \Rightarrow \Delta^\vee$ 4, 2, Obs2(d)
(6) $B \Rightarrow \Delta$ 5, possibly (($\bot$Intr) and (Cut)) or ($\vee\uparrow$)

(b) With respect to line (4) of the derivation below, we remark that $[A \vee B] \wedge C \Rightarrow [A \wedge C] \vee [B \wedge C]$ (Dist) can be proved in $\mathbf{\Sigma HP}$ in a way similar to the proof of Obs4(b) above. We procede then as follows.

(1) $\Gamma^\wedge \wedge \Theta^\wedge \Rightarrow \Delta^\vee \vee A$ Hyp., ($\wedge$Intr), ($\wedge\downarrow$), ($\vee\downarrow$)
(2) $\Gamma^\wedge \wedge \Theta^\wedge \Rightarrow \Theta^\wedge$ Obs2(a), (Intr$\wedge$)
(3) $\Gamma^\wedge \wedge \Theta^\wedge \Rightarrow [\Delta^\vee \vee A] \wedge \Theta^\wedge$ 1, 2, (Intr$\wedge$)
(4) $\Gamma^\wedge \wedge \Theta^\wedge \Rightarrow [\Delta^\vee \wedge \Theta^\wedge] \vee [A \wedge \Theta^\wedge]$ 3, Dist., Obs2(d)
(5) $\Delta^\vee \wedge \Theta^\wedge \Rightarrow \Delta^\vee \vee \Lambda^\vee$ Obs2(a), ($\wedge$Intr), (Intr$\vee$)

(6) $A \wedge \Theta^\wedge \Rightarrow \Delta^\vee \vee \Lambda^\vee$ Hyp., (Intr$\vee$)
(7) $[\Delta^\vee \wedge \Theta^\wedge] \vee [A \wedge \Theta^\wedge] \Rightarrow \Delta^\vee \vee \Lambda^\vee$ 5, 6, ($\vee$Intr)
(8) $\Gamma^\wedge \wedge \Theta^\wedge \Rightarrow \Delta^\vee \vee \Lambda^\vee$ 4, 7, Obs2(d)

As before, we use ($\wedge\uparrow$) and ($\vee\uparrow$) to convert the last sequent back into a sequent built up from formulas from Γ, Δ, Θ, and Λ. In doing so, we have possibly to employ (Intr$\top$) and ($\bot$Intr) as well as rule (PCut) from (a) in order to eliminate occurences of $\bot$ and $\top$.   ⊞

It is also possible now to derive Axiom A9 of **A**.

Observation 10 G_2 renders $\Rightarrow [\top \to A] \to [B \to A]$ derivable.   ⊟

Proof. (1) $\top, [\top \to A]^* \Rightarrow A$ Obs2(a), (Elim$\to$)
(2) $B \to \top, \top \to A \Rightarrow B \to A$ 1, ($\to\star$)
(3) $\top \to A \Rightarrow B \to A$ 2, Obs7(b), (PCut)
(4) $\Rightarrow [\top \to A] \to [B \to A]$ 3, Obs2(c)   ⊞

Above, A9 was called the "modified exportation principle" because of its relationship to the inference rule usually called "exportation" leading from a premiss $A \wedge B \to C$ to the conclusion $A \to [B \to C]$ by "exporting" the formula A out of the premiss's hypothesis. The idea of exportation is an important one having obvious connections with the deduction theorem, the algebraic idea of residuation, and the category theoretic idea of relating products (conjunctions) to exponentials (conditionals). But exportation offends both the doctrines of relevance logic[22] and Ackermann's homogenity principle. However, Ackermann (1950b, 43) uses A9 to derive a weakened but homogeneous form of exportation leading from $A \wedge B \to C$ to $[\top \to A] \to [B \to C]$. Let us give the special lable **ΣXP** to the "exportative" logics arising from **ΣHP** by adopting generalizations G_1, and G_2. The schema from Obs10 is a weakened and "homogenized" version of the notorious *verum ex quodlibet* $A \to [B \to A]$, one of the so-called "paradoxes of implication". Since the latter principle is inhomogeneous, it is not acceptable from Ackermann's point of view. However, we recognize by Obs10 that **A** contains "weakly paradoxical" principles as we have recognized by the same observation that it is also "weakly exportative".

Both intuitionistic and classical logic are even more "paradoxical" by adopting the full *verum ex quodlibet* $A \to [B \to A]$. This principal is easily derived when one gives up Ackermann's homogenity principle as implemented in ($\to\star$) and ($\star\to$). We retain the latter rule (as liberalized by G_2) but extend G_2 by modifying the restrictive conditions on ($\to\star$) as in (10) below.

---

[22] Cp. Anderson's and Belnap's discussion of the "Fallacies of exportation" in §22.2.2 of their Anderson and Belnap (1975).

Conditions: (10)

1. $|\Delta| = 1$,
2. At least one of the members of Δ gets the prefix if this is attached to at least one of the formulas in Γ.

We call this rule "$(\to\star_i)$". Given (SCut), (GCut), and $(\to\star_i)$, we are able to derive Genzen's (FES) (only for succedents consisting of only a single formula, of course, because of the first restriction on $(\to\star_i)$), see (3) above. Logics adopting these rules are thus "strongly exportative" by allowing to transport each formula from the antecedent to the succedent — not just those having been originally imported from there by (Elim$\to$). The star device becomes redundant and we may therefore cancel the stars in (Elim$\to$). We denote this modification of $\Sigma\mathbf{XP}$ by "G_3"; G_3 (together with G_1 and G_2, of course) yields the Σ-formulation $\Sigma\mathbf{IP}$ of intuitionistic logic.

Classical logic, besides being strongly exportative, is also "distributive" in the sense that it gives up the first restriction of (10) thus allowing for the possibly unhomgeneous distribution of the prefix $A \to$ over several formulas in the succedent Δ. Let us call this strengthening of G_3 "G_4" and the resulting liberalized rule "$(\to\star_k)$". Given $(\to\star_k)$, it is easy to derive Peirce's law, the hallmark of classical implication.

Observation 11 *In the system $\Sigma\mathbf{KP}$ resulting from $\Sigma\mathbf{HP}$ by G_1, G_2, and G_4, Peirce's law becomes derivable.* ⌐

Proof.
(1) $A \Rightarrow A, B$ Obs2(a), Obs8
(2) $\Rightarrow A, A \to B$ 1, $(\to\star_k)$, $(\to\text{Ref})$, (GCut)
(3) $\Rightarrow A \vee [A \to B]$ (2), $(\vee\downarrow)$
(4) $A \Rightarrow [[A \to B] \to A] \to A$ Obs2(a), $(\to\star_k)$
(5) $[A \to B] \to A, A \to B \Rightarrow A$ Obs2(a), (Elim$\to$), Obs4(a)
(6) $A \to B \Rightarrow [[A \to B] \to A] \to A$ 5, $(\to\star_k)$, Obs2(a), (GCut)
(7) $A \vee [A \to B] \Rightarrow [[A \to B] \to A] \to A$ (4), (6), ($\vee$Intr)
(8) $\Rightarrow [[A \to B] \to A] \to A$ (3), (7), (GCut)

⊞

As in the case of exportation, Ackermann's **A** does not implement the complete liberalization of G_4, but it follows classical logic a short part of way in the revision of $(\to\star)$. It retains, of course, the requirement of homogenity. Thus, if a condition is attached to one formula of the succedent, then every formula should get the same attachment there. Furthermore, only $\top$ is allowed as a condition to be attached when the succedent comprises two formulas. In this way, we arrive at the following conditions for the modified rule $(\to\star_a)$:

Conditions: (11)
1. $|\Delta| = 1$ unless $A = \top$.
2. Each unstarred formula of Γ and Δ is attached the prefix $A \to$.
3. No starred formula receives this prefix.

Let us call this weaker liberalization "G$_4'$" and the thus liberalized rule "($\to\star_a$)". Given ($\to\star_a$), the distribution axiom A10 of **A** follows imediately from the easily derivable $A \lor B \Rightarrow A, B$. Thus G$_4'$ leads up to the Σ-formulation Σ**AP** of the positive propositional fragment of Ackermann's **A**. — We conclude by giving a diagram depicting the relationships of the logics discussed.

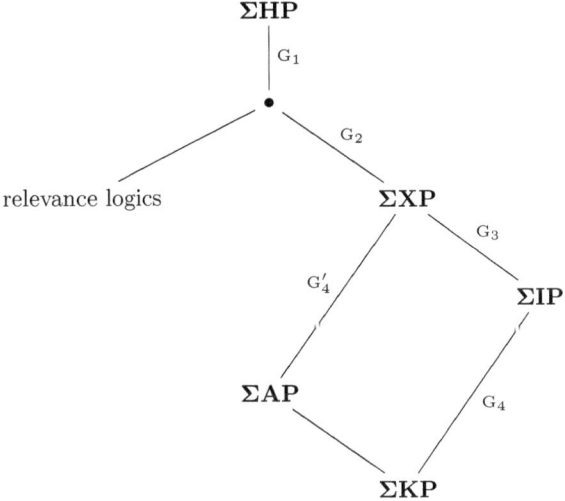

Fig. 2. Some Logics and Their Relation to Ackermann's System

5 Negation

Unlike Ackermann, we included the constant $\bot$ in our language. This makes it possible to define negation "in terms of absurdity" — a possibility briefly discussed by Ackermann (1950b, 39).

$$\neg A \underset{\text{def}}{\Leftrightarrow} A \to \bot \qquad (12)$$

The thus defined connective may serve as negation within the framework of intuitionistic and classical logic. However, there are two objections against introducing negation in this way into Ackermann's **A**. First, it places the negation of a proposition one level higher in the semantic hierarchy than the proposition itself. Secondly, being absurd, *i.e.* implying a contradiction, does not mean the same as being false. For the Russellian class R, the proposition R ∈ R implies a contradiction. But according to Behmann's analysis (see sec. 3 above) this proposition is not false, rather it lacks a truth value.

In partial logic falsity is treated as a notion independent from and on a par with truth. This is another corner stone of Ackermann's system; cp. the introduction of Ackermann (1952b). Consequently, we should supplement the truth conditions stated by the rules of the connective module by corresponding falsity conditions. Such conditions should be formulated also for conditionals and for negations, of course. Furthermore, we should add a principle encoding Ackermann's view that falsehood implies absurdity.[23] All these rules make up the "negation module" displayed in Table 4. We will denote a system resulting from **ΣHP**, **ΣXP**, **ΣIP**, or **ΣKP** by supplying the system with the negation module by adding the letter "N" to the respective systems' name, *e.g.*, **ΣHPN**, *etc.*

Ackermann's propositional axioms characterizing negation are easily derivable already in our base system **ΣHPN**.

Observation 12 *The propositional negation axioms A12 to A18b of* **A** *are derivable in* **ΣHPN**. ⊟

Proof. The relevant derivations are quite elementary. We give thus only two examples.

A13b (1) ⊤ ⇒ ∼⊥ (Intr∼⊥), (⊤Intr)
     (2) ⇒ ⊤ → ∼⊥ 1, Obs2(c)

A16a (1) ∼A ⇒ ∼A ∨ ∼B Obs2(a), (Intr∨)
     (2) ∼B ⇒ ∼A ∨ ∼B analogously
     (3) ∼[A ∧ B] ⇒ ∼A ∨ ∼B 1, 2, (∼∧Intr)
     (4) ⇒ ∼[A ∧ B] → ∼A ∨ ∼B 3, Obs2(c)

⊞

Translation of sequents into formulas again transforms the rules of the negation module to derviable rules of **A**. **ΣAPN** thus turns out to be equivalent with the propositional fragment of the Hilbert system **A**. — Given the difference in semantic level between ¬A and ∼A, it is not so easy to compare the two negations within **ΣAPN**. The following observation states that, in intuitionistic logic, ∼ is really stronger than ¬ and that the distinction between the two negations collapses, as expected, in **ΣKPN**.

---

[23] "Die Falschheit einer Aussage impliziert zwar die Absurdität, sagt aber im allgemeinen doch mehr" (Ackermann, 1950b, 39).

Table 4. Falsity Conditions for the Logical Connectives

The negation module

(excon) $A, \sim A \Rightarrow \bot$

($\sim\top$Intr) $\sim\top \Rightarrow$

(Intr$\sim\top$) $\dfrac{\Gamma \Rightarrow \Delta}{\Gamma \Rightarrow \Delta, \sim\top}$

($\sim\bot$Intr) $\dfrac{\Gamma \Rightarrow \Delta}{\sim\bot, \Gamma \Rightarrow \Delta}$

(Intr$\sim\bot$) $\Rightarrow \sim\bot$

($\sim\sim$Intr) $\dfrac{A, \Gamma \Rightarrow \Delta}{\sim\sim A, \Gamma \Rightarrow \Delta}$

(Intr$\sim\sim$) $\dfrac{\Gamma \Rightarrow \Delta, A}{\Gamma \Rightarrow \Delta, \sim\sim A}$

($\sim\wedge$Intr) $\dfrac{\sim A, \Gamma \Rightarrow \Delta \quad \sim B, \Gamma \Rightarrow \Delta}{\sim[A \wedge B], \Gamma \Rightarrow \Delta}$

(Intr$\sim\wedge$) $\dfrac{\Gamma \Rightarrow \Delta, \sim A}{\Gamma \Rightarrow \Delta, \sim[A \wedge B]}$

(Intr$\sim\wedge$) $\dfrac{\Gamma \Rightarrow \Delta, \sim B}{\Gamma \Rightarrow \Delta, \sim[A \wedge B]}$

($\sim\vee$Intr) $\dfrac{\sim A, \Gamma \Rightarrow \Delta}{\sim[A \vee B], \Gamma \Rightarrow \Delta}$

(Intr$\sim\vee$) $\dfrac{\Gamma \Rightarrow \Delta, \sim A \quad \Gamma \Rightarrow \Delta, \sim B}{\Gamma \Rightarrow \Delta, \sim[A \vee B]}$

($\sim\vee$Intr) $\dfrac{\sim B, \Gamma \Rightarrow \Delta}{\sim[A \vee B], \Gamma \Rightarrow \Delta}$

($\sim\to$Intr) $\dfrac{\top \to A, B \to \bot, \Gamma \Rightarrow \Delta}{\sim[A \to B], \Gamma \Rightarrow \Delta}$

(Intr$\sim\to$) $\dfrac{\Gamma \Rightarrow \Delta, \top \to A \quad \Gamma \Rightarrow \Delta, B \to \bot}{\Gamma \Rightarrow \Delta, \sim[A \to B]}$

Observation 13 *We have:*

(a) in Σ**IPN**: $\sim A \Rightarrow \neg A$.
(b) in Σ**KPN**, *also conversely,* $\neg A \Rightarrow \sim A$. □

Proof. (a) This follows imediately by applying ($\to\star_i$) to (excon).

(b) Using ($\to\star_k$), we easily derive $\Rightarrow A, A \to \bot$ from $A \Rightarrow A, \bot$. Combining this with (excon) by cut, we get $\sim A \Rightarrow \bot, A \to \bot$. From this, we may eliminate the occurence of $\bot$ in the succedent by ($\bot$Intr) and cut (and an easily derived commutation in the succedent). ⊞

The idea to characterize a strong negation $\sim$ by rules as those shown in Table 4 is also known from the systems of Fitch — cp., for instance, Fitch (1952, 53–63) — and the constructive logic of Nelson and Markov; cp. (Rasiowa, 1974, chap. XII) as well as Wansing (1993) for the latter systems. Σ**APN**, therefore, bears a certain resemblance to these systems. The difference, however, between these logics and Ackermann's consists in the treatment of the conditional. The systems quoted use intuitionistic implication, which differs from Ackermann's conditional. Furthermore, the negation of implications is treated in different ways. Fitch is not concerned with falsity conditions for implication at all. In Nelson's and Markov's logic, on the other hand, $\sim[A \to B]$ is equivalent to $A \wedge \sim B$. Clearly, this is at odds with Ackermann's homogenity principle since $A \wedge \sim B$ is one semantic level below its (strong) negation $\sim[A \to B]$. Therefore, Ackermann declared $\sim[A \to B]$ to be equivalent to $[\top \to B] \wedge [B \to \bot]$ belonging to the same semantic level.

6 Quantification

Ackermann introduces ("in bekannter Weise") universal and existential quantification into his system as generalizations of conjunction and disjunction (cp. Ackermann, 1950b, 39f). A universal statement is a conjunction with infinitely many conjuncts and, dually, an existential statement is a disjunction with infinitely many disjuncts. This poses a problem for a sequent system in which conjunction and disjunction are — *via* rules like $(\wedge\!\Uparrow)$ and $(\vee\!\Uparrow)$ — reflections of the structural comma-operation (concatenation). Given the thus described interpretation of conjunction and disjunction, the most natural thing to do for implementing Ackermann's generalization would perhaps be to admit sequents with infitely many formulas in the antecedent and the succedent as the purely structural counterparts of quantification. This, of course, would not lead up to a calculus in the standard understanding of this notion. We will instead simply adopt a standard rule set for quantification to which we add the obvious clauses introducing negations of quantified statements; cp. Table 5. It should be remembered that Ackermann distinguishes typographically between bound (x, y, z) and free (a, b, c) variables. In the following Table 5 the term "Quantifier Condition" refers to the requirement that the free variable a should not occur in Γ, Δ.

The procedure taken has its drawbacks, however. Sequents with "finitary" conjunction and disjunction have completely structural counterparts. Thus, after having explained quite generally for arbitrary formulas within sequents how they may be modified by means of the conditional connective in order to derive a new sequent, we can conclude for the special cases of conjunctions and disjunctions how the conditional interacts with them by making use of the latter's structural analogues. An example of this strategy is the proof of Obs3 above. Given that universal and existential quantification unlike their finitary counterparts do not possess completely structural analogues, this important proof strategy breaks down in the case of quantification. For the weaker ones of our systems (those weaker than the intuitionistic system), we have therefore to compensate for this loss by modifying the rules $(\star \to)$ and $(\to \star)$ in such a way that formulas $\forall x.A(x)$ (in the antecedent) and $\exists x.A(x)$ (in the succedent) may be interpreted as infinitary connectives in the way alluded to by Ackermann. For this purpose, we introduce two new pieces of notation. First, $\Delta^\star \to A$ is used to denote any sequence arising from Δ by suffixing the latter sequence's formulas either by $\to A$ or alternatively (but not at all obligatory) changing

Table 5. The Quantification Module

The quantification module

$(\forall\text{Intr}) \dfrac{A(a), \Gamma \Rightarrow \Delta}{\forall x.A(x), \Gamma \Rightarrow \Delta}$
$\qquad (\text{Intr}\forall) \dfrac{\Gamma \Rightarrow \Delta, A(a)}{\Gamma \Rightarrow \Delta, \forall x.A(x)}$

Quantifier Condition

$(\sim\forall\text{Intr}) \dfrac{\neg A(a), \Gamma \Rightarrow \Delta}{\neg\forall x.A(x), \Gamma \Rightarrow \Delta}$
$\qquad (\text{Intr}\sim\forall) \dfrac{\Gamma \Rightarrow \Delta, \neg A(a)}{\Gamma \Rightarrow \Delta, \neg\forall x.A(x)}$

Quantifier Condition

$(\exists\text{Intr}) \dfrac{A(a), \Gamma \Rightarrow \Delta}{\exists x.A(x), \Gamma \Rightarrow \Delta}$
$\qquad (\exists\text{Intr}) \dfrac{\Gamma \Rightarrow \Delta, A(a)}{\Gamma \Rightarrow \Delta, \exists x.A(x)}$

Quantifier Condition

$(\sim\exists\text{Intr}) \dfrac{A(a), \Gamma \Rightarrow \Gamma}{\neg\exists x.A(x), \Gamma \Rightarrow \Delta}$
$\qquad (\text{Intr}\sim\exists) \dfrac{\Gamma \Rightarrow \Delta, A(a)}{\Gamma \Rightarrow \Delta, \neg\exists x.A(x)}$

Quantifier Condition

---

a member $\exists x.B(x)$ to $\forall x.[B(x) \to A]$. The second operation comes along in two variants: $\to^\star_\forall$ and $\to^\star_\exists$. We describe them simultaneously by using the notatiton $\to^\star_Q$. $(A \to^\star_Q)\Gamma$ denotes any sequence of formulas resulting from Γ by adjoining to no, some, or all members of Γ either the prefix $A \to$ or, as an allowed (but again not obligatory) alternative in case of a member $Qx.B$, by adding the prefix to the matrix B thus yielding $Qx.[A \to B]$. The schemes for the rules $(\star\to)$ and $(\to\star)$, then, look like this:

$$(\to\star) \dfrac{\Gamma \Rightarrow \Delta}{(A \to^\star_\forall)\Gamma \Rightarrow (A \to^\star_\exists)\Delta} \qquad (\star\to) \dfrac{\Gamma \Rightarrow \Delta}{\Delta^\star \to A \Rightarrow \Gamma \to A} \qquad (13)$$

If in an application of $(\to\star)$, the increment $A \to$ is inserted into a formula rather than prefixed to it, we call this application "insertive". Furthermore, within Ackermann's system, a succedent Δ is called "multiple" if $|\Delta| > 1$ or if Δ consists of an existential formula $\exists x.A$. We are now going to define the quantificational counterparts **ΣHQN**, **ΣXQN**, **ΣIQN**, **ΣAQN**, and **ΣKQN** of the propositional systems **ΣHPN**, **ΣXPN**, **ΣIPN**, **ΣAPN**, and **ΣKPN**. In all cases, of course, we shift to the language of predicate logic. In the cases of **ΣIPN** and **ΣKPN**, the only further modification consists in adding the rules of the quantificational module. For the remaining systems, the rules $(\to\star)$ and $(\star\to)$ are modified by adopting the schemes of (13). In **ΣHQN** and **ΣXQN**, we add to the restrictions of the rule $(\to\star_a)$ the further condition:

An application of this rule may not be insertive. In $\mathbf{\Sigma AQN}$ we replace the first restriction of (11) by: "Δ may not be multiple unless $A = \top$".

Observation 14 *As regards quantification,*

(a) the set of derivable formulas in all systems mentioned are closed under universal generalization.

(b) all instances of axioms A20–A25 and A27a–A28b are derivable in all the systems mentioned;

(c) furthermore, we have in the classical system $\mathbf{\Sigma KQN}$ that $[A \to \exists x.B(x)] \to \exists x.[A \to B(x)]$;

(d) in $\mathbf{\Sigma AQN}$ all instance of the latter scheme are derivable where A is $\top$; thus we have A25 in that system. □

Proof. (a) From $\Rightarrow A(a)$ we conclude $\Rightarrow \forall x.A(x)$ by a simple application of (Intr∀).

(b) The proofs are rather obvious. Therefore only two examples are given. In the case of A21, we apply the appropriate version of the modified rule $(\to\star)$ to $\forall x.B(x) \Rightarrow \forall x.B(x)$ (from Obs2a) in order to get $\forall x.[A \to B(x)] \Rightarrow A \to \forall x.B(x)$. To this, Obs2(c) is applied. — For A23, we start from $\exists x.A(x) \Rightarrow \exists x.A(x)$. Applying the appropriate version of the extended rule $(\star\to)$ and then Obs2(c) yields the desired A23.

(c) Note that we have $\Rightarrow A, A \to B(a)$ in $\mathbf{\Sigma KQN}$; this is derived as line (3) in our above proof of Peirce's law (see Obs11). By (∃Intr), we get $\Rightarrow A, \exists x.[A \to B(x)]$. On the other hand, we have $A, A \to \exists x.[A \to B(x)] \Rightarrow \exists x.[A \to B(x)]$ by an application of (Elim→) to an instance of Obs2a. Combining the two sequents by cut (Obs9b), we get $A \to \exists x.[A \to B(x)] \Rightarrow \exists x.[A \to B(x)]$. Denote this sequent by (•). We procede then as follows:

    (1) $B(a) \Rightarrow A \to B(a)$ Obs2a, $(\to\star_k)$
    (2) $B(a) \Rightarrow \exists x.[A \to B(x)]$ 1, (Intr∃)
    (3) $\exists x.B(x) \Rightarrow \exists x.[A \to B(x)]$ 2, (∃Intr)
    (4) $A \to \exists x.B(x) \Rightarrow A \to \exists x.[A \to B(x)]$ 3, $(\to\star_k)$
    (5) $A \to \exists x.B(x) \Rightarrow \exists x.[A \to B(x)]$ by cut from 4 and (•)
    (6) $\Rightarrow \exists x.[A \to B(x)] \to \exists x.[A \to B(x)]$ 5, Obs2c

(d) In the proof of (c), we relied on the liberal rule $(\to\star_k)$. Of course, this rule is not availabel in $\mathbf{\Sigma AQN}$. A26, however, follows by a simple application of the extended rule $(\to\star_a)$ to $\exists x.A(x) \Rightarrow \exists x.A(x)$. ⊞

Summing up, we conclude that all quantificational theorems of $\mathbf{A}$ are derivable in $\mathbf{\Sigma AQN}$. Conversely, it is not difficult to show that the "formula translations" of the quantificational rules are derivable in $\mathbf{A}$ and the modified rules for the conditional are derivable rules of the latter systems. Thus the quantificational fragment of $\mathbf{A}$ and $\mathbf{\Sigma AQN}$ are equivalent.

7 Class Theory

Starting from its quantificational fragment, we reach at Ackermann's class theory by adding principles for the assertion and denial of identity and membership statements. The identity principles A28–A30 are rather standard and the class logical principles A31a–AA32b amount to unrestricted comprehension for classes and relations. Extensionality, however, is missing. It is rather obvious how to encode the identity and comprehension principles within the Σ-framework. Therefore, instead of stating corresponding rules, we finish this article by considering the fate of the two paradoxes from which we started in order to motivate Ackermann's class theory. Ackermann (1950b, sec. 3) proves[24] the consistency of his system by a transfinite induction up to ω^{ω^4}. Therefore we may be assured that the paradoxes do not arise in **A**. Nevertheless, it is instructive to see where the usual arguments leading to the paradoxes break down within his framework.

Having unrestricted abstraction at our disposal, it is easy to derive both $R \notin R \to R \in R$ (I) and its converse $R \in R \to R \notin R$ (II; remember that R is Russell's class). Frege (1893/1903, II, 256f) gives two derivations of Russell's paradox within his system, both making use of the principle $[A \to \sim A] \to \sim A$ (F), which Frege denotes by "Ig". Of course, (I) and (II) together with this principle lead up to a contradiction by simple applications of *modus ponens*. As is clear from Ackermann's consistency proof, the inhomogeneous principle (F) is not derivable within **A**. Starting, however, from (II) and the trivial $R \in R \to R \in R$ we derive by means of A4 and A13 that $\neg R \in R$, where the weak negation connective $\neg$ is defined as above in (12). In a similar way, we conclude $\neg R \notin R$ (*i.e.*, $\sim R \subset R$). Thus both the Russellian proposition $R \in R$ as well its strong negation $\sim R \in R$ are absurd (in the technical sense of implying $\bot$). As it should be in such a case, none of them is provable.

Curry's paradox — as relying on the argument (2) above — is avoided since the deduction theorem breaks down in Ackermann's system. Though **A** is "weakly exportative" (cp. p. 45 above for the discussion of A9), it does neither allow to infer $A \to [B \to C]$ from $A \wedge B \to C$ nor to conclude, on the metalevel[25], $M \vdash A \to B$ from $A, M \vdash B$. Therfore, the transition from lines 1 and 3 to 4 in the above argument (2) leading up to Curry's paradox is not valid in Ackermann's **A**. However, there is an alternative version of Curry's paradox which uses the contraction rule (Con)[26] instead of the deduction theorem.

$$\text{Con } \frac{A \to [A \to B]}{A \to B} \qquad (14)$$

---

[24] Harrop (1954, 497) hints at a minor oversight in Ackermann's original proof which can easily be corrected. Ackermann has given some modifications of his original argument in the articles following the publication of his (1950b).

[25] Weak exportativity is, however, reflected by the validity of a weakened deduction theorem; cp. Ackermann (1950b, 45) and Harrop (1954, 504).

[26] To block contraction is the idea underlying the class theory in Grishin (74).

to step from line (2) of the above derivation to line (3). Of course, we know that the rule (Con) cannot be derivable in **A** and its premiss is suspect from the viewpoint of Ackermann's homogenity principle. Nevertheless, there are rules derivable in **A** having premisses not compatible with this principle; cp. rule VII in Ackermann (1950b, 43). Furthermore, (Con) would be a derivable rule — and thus would give way to Curry's paradox — if the "formula version" $A \wedge [A \rightarrow B] \rightarrow B$ of *modus ponens* were derivable[27]. In this case, we would conclude from $A \rightarrow [A \rightarrow B]$ and the trivial $A \rightarrow A$ that $A \rightarrow [A \wedge [A \rightarrow B]]$ and from this, by using the linear form of *modus ponens*, $A \rightarrow B$. Again, we know from the consistency proof that $A \wedge [A \rightarrow B] \rightarrow B$ is not derivable. A matrix argument of the usual kind for the independence of this formula can be found in Harrop (1954, 511).

References

Wilhelm Ackermann. *Ein System der typenfreien Logkik*. Hirzel, Leipzig, 1941.

Wilhelm Ackermann. Review of Halldén (1949). *The journal of symbolic logic*, 15:225–226, 1950a.

Wilhelm Ackermann. Widerspruchsfreier Aufbau der Logik, I. *The journal of symbolic logic*, 15:33–57, 1950b.

Wilhelm Ackermann. Review of Fitch (1952). *The journal of symbolic logic*, 17:266–268, 1952a.

Wilhelm Ackermann. Widerspruchsfreier Aufbau einer typenfreien Logik (Erweitertes System). *Mathematische Zeitschrift*, 55:364–384, 1952b.

Wilhelm Ackermann. Widerspruchsfreier Aufbau einer typenfreien Logik, II. *Mathematische Zeitschrift*, 57:155–166, 1953.

Wilhelm Ackermann. Begründung einer strengen Implikation. *The journal of symbolic logic*, 21:113–128, 1956a.

Wilhelm Ackermann. Zur Axiomatik der Mengenlehre. *Mathematische Annalen*, 131:336–345, 1956b.

Wilhelm Ackermann. Ein typenfreies System der Logik mit ausreichender mathematischer Anwendungsfähigkeit, I. *Archiv für mathematische Logik und Grundlagenforschung*, 4:3–26, 1958.

Wilhelm Ackermann. Ein typenfreies System der Logik mit ausreichender mathematischer Anwendungsfähigkeit, II (Ableitung der Zahlentheorie). *Archiv für mathematische Logik und Grundlagenforschung*, pages 96–111, 1959.

---

[27] Note again the connection with the deduction theorem. Its strong version might be formulated thus: $A, M \vdash B$ iff $M \vdash A \rightarrow B$. Applying this (from right to left) to $A \rightarrow B \vdash A \rightarrow B$, we get *modus ponens*. The latter rule may thus be conceived as the converse of the (weak) deduction theorem, and the strong deduction theorem as the conjunction of the weak one and *modus ponens*.

Wilhelm Ackermann. Der Aufbau einer höheren Logik. *Archiv für mathematische Logik und Grundlagenforschung*, 7:5–22, 1965.

Alan Ross Anderson and Nuel D. Belnap. *Entailment. The logic of relevance and necessity*, volume I. Princeton University Press NJ, Princeton, 1975.

Alan Ross Anderson, Nuel D. Belnap, and Michael J. Dunn. *Entailment. The logic of relevance and necessity*, volume II. Princeton University Press, Princeton NJ, 1992.

Heinrich Behmann. Zu den Widersprüchen der Logik und Mengenlehre. *Jahresbericht der Deutschen Mathematiker-Vereinigung*, 40:37–41, 1931a.

Heinrich Behmann. Zuschrift an die Herausgeber: Zur Richtigstellung einer Kritik meiner Auflösung der logisch-mengentheoretischen Widersprüche. *Erkenntnis*, 2:305–306, 1931b.

Georg Cantor. Beiträge zur Begründung der transfiniten Mengenlehre. *Mathematische Annalen*, 46/49:481–512/207–246, 1895/1897. Reprinted: Cantor (1932, 282-356).

Georg Cantor. *Gesammelte Abhandlungen mathematischen und philosophischen Inhalts. Mit erläuternden Anmerkungen sowie mit Ergänzungen aus dem Briefwechsel Cantor-Dedekind*. Ed. by Ernst Zermelo. Springer, Berlin, 1932. Reprinted: Olms: Hildesheim 1962, 1980; Springer: Berlin *etc.* 1980.

Alonzo Church. A set of postulates for the foundations of logic, I. *Annals of mathematics (Series 2)*, 33:346–366, 1932.

Haskell B. Curry. The inconsistency of certain formal logics. *The journal of symbolic logic*, 7:115–117, 1942.

Kosta Došen. A historical introduction to substructural logics. In Kosta Došen and Peter Schroeder-Heister, editors, *Substructural logics*, pages 1–30. Clarendon Press, 1993.

Solomon Feferman. Towards useful type-free theories, I. *The journal of symbolic logic*, 49:75–111, 1982.

Frederic Brenton Fitch. An extension of basic logic. *The logic of symbolic logic*, 13:95–106, 1948.

Frederic Brenton Fitch. *Symbolic logic*. Ronald Press, New York, 1952.

Robert C. Flagg and John Myhill. Implication and analysis in classical Frege structures. *Annals of pure and applied logic*, 34:33–85, 1987.

Gottlob Frege. *Grundgesetze der Arithmetik*, 2 vols. Pohle, Jena, 1893/1903. Reprinted: Olms, Hildesheim, 1962, 1966 and 1998.

Gerhard Gentzen. Untersuchungen über das logische Schließen. *Mathematische Zeitschrift*, 39:176–210 and 405–431, 1934/35.

Kurt Gödel. Russell's mathematical logic. In Paul A. Schilpp, editor, *The philosophy of Bertrand Russell*, pages 123–153. Northwestern University Press, Evanston, 1944. Reprinted: (Gödel, 1990, 119-141).

Kurt Gödel. What is Cantor's continuum problem? *American mathematical monthly*, 54:515–525, *errata* vol. 55, 151, 1947. Reprinted: Gödel (1990, 176-187).

Kurt Gödel. *Collected works*, volume II: Publications 1938-1974. Oxford University Press, New York and Oxford, 1990.

Vyacheslav Nikolaevich Grishin. A non-standard logic and its application to set theory [in Russian]. In *Studies in formalized languages and nonclassical logics*, pages 135–171. Nauka, Moscow, 74.

Sören Halldén. *The logic of nonsense*. Uppsala Universitets Årskrift 1949: 9. Lundequistska Bokhandlen, Uppsala, 1949.

Ronald Harrop. An investiagtion of the propositional calculus used in a particular system of logic. *Proceedings of the Cambridge Philosophical Society*, 50:495–512, 1954.

Hans Hermes. Wilhelm Ackermann 1896–1962. *Notre Dame journal of formal logic*, 8:1–8, 1967.

Oiva Ketonen. *Untersuchungen zum Prädikatenkalkül*. Annales Academiae Scientiarum Fennicae. Series A. I. Mathematica-Physica. Vol. 23. Suomalainen Tiedeakatemia, 1944.

Stanisław Leśniewski. Über Definitionen in der sogenannten Theorie der Deduktion. *Comptes rendus de Séances de la Societè des Sciences et des Lettres de Varsovie, Cl. iii*, 24:289–309, 1931.

John Myhill. Levels of implication. In Alan Ross Anderson, Ruth Barcan Marcus, and Richard M. Martin, editors, *The logical enterprise*, pages 179–185. Yale University Press, New Haven, 1975.

John Myhill. Paradoxes. *Synthese*, 60:129–143, 1984.

Helena Rasiowa. *An algebraic approach to non-classical logics*. North-Holland Publishing Company, 1974.

William Reinhardt. Ackermann's set theory equals ZF. *Annals of mathematical logic*, 2:189–249, 1970.

Arnold Schmidt. *Mathematische Gesetze der Logik*, volume I. Springer, Berlin, 1960.

Christian Thiel. Gödels Anteil am Streit über Behmanns Behandlung der Antinomien. In Eckehart Köhler *et al.*, editor, *Wahrheit und Beweisbarkeit. Leben und Werk Kurt Gödels*, volume 2, pages 387–394. Öbv & Hpt, Wien, 2002.

Mary Tiles. *The philosophy of set theory: an introduction to Cantor's paradise*. Blackwell, Oxford, 1989.

Ellen Walther-Klaus. *Inhalt und Umfang. Untersuchungen zur Geltung und zur Geschichte der Reziprozität von Extension und Intension*. Olms, Hildesheim *etc.*, 1987.

Heinrich Wansing. *The logic of information structures*. Springer, Berlin, 1993.

Nominalistic Logic: From Naive Set Theory to Intensional Type Theory

Jørgen Villadsen

Department of Informatics and Mathematical Modeling
Technical University of Denmark

jv@imm.dtu.dk

Summary. Paul C. Gilmore presented Intensional Type Theory (ITT) in 2001 as a foundation of mathematics without non-logical axioms. This claim was stressed in his monograph *Logicism Renewed* (2005), but the split between logical and non-logical axioms is problematic, and furthermore the claim distracts from the main novelty: a type system that allows more constructions from axiomatic set theory than the well-known simple type theory does.

Nominalistic Logic (NL) is a new presentation of ITT as a sequent calculus together with a succinct nominalization axiom (N) that permits names of predicates as individuals in certain cases. Essentially NL has a flexible comprehension axiom, but no extensionality axiom and no infinity axiom, although axiom N is the key to the derivation of Peano's postulates for the natural numbers.

It is found that NL throws light on the principles of mathematics as understood by Russell in 1901 when discovering the paradox that characterizes naive set theory. Furthermore NL is shown to relate to seminal work in logic and computer science around 1950 and again around 2000, and with its unique features as a higher order logic, it seems particularly useful for advanced computer systems and tools with integrated mathematics.

> Computers are consummate nominalists.
>
> Gilmore — Page x of *Logicism Renewed* (Gilmore, 2005).

1 Introduction

In the following we discuss a number of issues in relation to the foundations of mathematics from the point of view of computer science. However the discussion also involves many historical and methodological investigations. The focus is on type theory and its relation to set theory.

In fact it seems that the entire aim and scope of the foundations of mathematics are very much dependent on the time in history and especially on the technological environment. This is in contrast to a view of mathematics as a purely abstract science. In particular, that which is considered the appropriate standard foundation of mathematics at one time in history, need not be the most fruitful approach at another time. For this reason we do not start with a presentation of Nominalistic Logic (NL) but instead start with an investigation of a more historical and methodological nature. Furthermore, discussion of a series of other logics will also make discussion of NL easier.

1.1 Overview

It was Frege who in Frege (1879) took the first and crucial steps in "modern" symbolic logic. In 1893 Frege published the first of two books on THE BASIC LAWS OF ARITHMETIC (1893 and 1903) but his book was little studied at that time. Russell did most of the work on the THE PRINCIPLES OF MATHEMATICS in 1900 and while considering the work of Cantor he discovered in 1901 the contradiction that is now called Russell's paradox (Grattan-Guinness, 2000, page 311). In 1902 Russell saw that the paradox could also be formulated in the work of Frege, and after completing an appendix on the work of Frege and another appendix with the first thoughts about a theory of types in order to avoid the formulation of the paradox, he sent a letter to Frege pointing out the inconsistency. Frege had just completed the main part of his second book on THE BASIC LAWS OF ARITHMETIC but added an appendix to the book with a discussion of the paradox in 1903. Actually Zermelo discovered the paradox[1] in 1899 but he did not publish it (Grattan-Guinness, 2000, page 216) and in 1908 he was even the first to call it Russell's paradox.

It turns out that the 100 years between Russell's discovery in 1901 of the paradox that characterizes naive set theory and Gilmore's publication in 2001 of Intentional Type Theory (ITT) can be split in the middle according to the following table:

... –1900	Axiomatization of Mathematics
1901–1950	Formalization of Mathematics
1951–2000	Mechanization of Mathematics
2001–...	Integration of Mathematics

Of course the splits 1900/1901, 1950/1951 and 2000/2001 should not be taken as exact dates but we hope to show that the periods are quite distinct with respect to the work on the foundations of mathematics. — The notions of axiomatized, formalized, mechanized and integrated mathematics are explained in sections 2, 3, 4 and 5, respectively, but we shall also briefly point out the main features here.

---

[1] Cp. the contribution of Peckhaus to this volume, p. 11–19.

Nominalistic Logic: From Naive Set Theory to Intensional Type Theory   59

The axiomatic method has a long history but the use of formal systems as understood today was rare before 1900. Axioms were discussed but almost always little attention was paid to the logic. Frege's system from 1879 was the main exception.

The period 1901–1950 is characterized by a change from the axiomatization of mathematics with an implicit logic to the formalization of mathematics with an explicit logic. Russell's paradox was an important discovery and marks the start of the period. Gödel's theorem from 1931 was a key result. The close connections between (symbolic) logic and (theoretical) computer science gradually became clear from 1936 and on. The idea of a logic borrows concepts from the idea of a machine — where a machine is a device that performs a task and not a device that consumes/produces energy — cf. the machine age with its hallmark of Henry Ford's automobile Model T and the idea of assembly line production from 1913. However the impact of logic on mathematics itself was rather limited. Seminal work in logic and computer science appeared around 1950 and is discussed later.

The period 1951-2000 is the computer age, where information technology had a huge impact on daily life and science. The impact of logic on mathematics itself was noticeable. Various formal systems providing foundations of mathematics were turned into computer programs yielding the mechanization of mathematics. It is important to note that this is by no means a simple task. Many well-known theories and theorems have been formalized and mechanized.

Finally we think that the current period is the internet age — not just the internet as we know it, but a more ubiquitous internet — with an even bigger impact on mathematics than the impact of the previous periods (the machine age and the computer age). Logics and foundations of mathematics will be integrated in various systems and this will have a huge impact on mathematics since mathematics is very much a social activity. There are also problems that are virtually impossible to solve without computers. For example, Thomas Hales produced in 1998 a computer-aided proof of the Kepler Conjecture (the most efficient way to pack spheres is in a pyramid shape). Then, in 1999, a panel of 12 referees was established in order to verify the correctness of the proof. After four full years the referees said that they were 99% certain of the correctness of the proof, but that they had been unable to completely verify the proof. Hales started the so-called Flyspeck project that seeks to formalize the proof of the Kepler Conjecture in the computer theorem prover HOL Light. The project involves many people and the integration of several computer tools and foundations of mathematics.

1.2 Set Theory versus Type Theory

In the following sections some basic familiarity with logic, set theory, type theory and the λ-calculus is assumed. See for example HANDBOOK OF MATH-

EMATICAL LOGIC (1977) edited by Jon Barwise. But before we go on we need to clarify some terminology.

By set theory, we mean (formal) axiomatic set theory. In the metalanguage, we will use the notion of sets in the usual informal way, but set theory is always a formal system (the use of a metalanguage with sets is a convenience only — other approaches with sufficient power are available, say, map theory; cf. Grue's contribution to this volume, p. 86–110). And unless otherwise noted the axioms will be assumed to be in first order logic.

The binary relation $=$ for equality is assumed. Alternatively $x = y$ stands for $\forall z. x \in z \to y \in z$. We shall later briefly describe first order logic without equality, called elementary logic (ℓ), to which the following equality axioms must be added in the presence of set theory with a specific binary relation $\in$ for membership:

$$x = x$$
$$x = y \to y = x$$
$$x = y \land y = z \to x = z$$
$$x_1 = x_2 \land y_1 = y_2 \to (x_1 \in y_1 \to x_2 \in y_2)$$

Naive set theory is a general way to refer to any set theory where Russell's paradox is possible. The first mention of naive set theory appears to be in a review by Hermann Weyl from 1946 in AMERICAN MATHEMATICAL MONTHLY (Weyl, 1946), while the following precise description by Hao Wang appeared in THE FORMALIZATION OF MATHEMATICS, a long essay from 1954 in JOURNAL OF SYMBOLIC LOGIC.

> At the beginning of the century the discovery of paradoxes plus the popularity of axiomatic method in geometry and arithmetic led quite naturally to attempts to construct axiom systems for set theory in which as much of Cantor's "naive" theory as possible, but of course none of the paradoxes, is to be derived. (Wang, 1954, 246)

The quotes around the word "naive" show that the common meaning of naive set theory as the informal development of set theory by Georg Cantor had not yet been completely established. Naive set theory, with or without quotes, was discussed by several other logicians in the period 1946–59. Then in 1960 Paul Halmos published a book entitled NAIVE SET THEORY, but the choice of title was unfortunate, because it dealt with the axiomatic set theory in which the Russell's paradox was avoided.

The problem in naive set theory relates to the formation of sets. One's first intuition might be that any sets can be formed, but this view leads to inconsistencies. The problem can be seen from the following theory in first order logic:

$$\exists y \forall x. x \in y \leftrightarrow \varphi$$

The variable y must not occur free in the formula φ (but the variable x can occur free). Usually it is called the comprehension axiom. It is really a set of axioms, given by a schema, but we do not bother to make the distinction here.

Russell's paradox is obtained by taking φ to be $\neg x \in x$ and then instantiating x to y.

$$\exists y. y \in y \leftrightarrow \neg y \in y$$

Note that the following extensionality axiom plays no role in the derivation of the paradox.

$$\forall xy.(\forall z.\ z \in x \leftrightarrow z \in y) \rightarrow x = y$$

Russell's paradox is blocked in axiomatic set theory as described by Joseph Shoenfield (cp. Barwise, 1977). More precisely, this is the set theory pioneered by Zermelo in 1908 and later improved by Fraenkel and Skolem around 1922. It is usually referred to as Zermelo-Fraenkel set theory and at least since a paper by J. Barkley Rosser and Hao Wang in 1950 it is denoted by "ZF" (see Rosser and Wang, 1950, 128) or "ZFC", the latter when the axiom of choice is added (shown to be independent of the other axioms by Paul Cohen in 1963).

In ZFC set theory the paradox is avoided by using a subset axiom, the so-called "axiom of separation".

$$\forall z \exists y \forall x. x \in y \leftrightarrow (x \in z \wedge \varphi)$$

As before the variable y must not occur free in the formula φ. But the problem is that it is a subset axiom and not a "proper" comprehension axiom. — Usually axiomatic set theory requires some kind of infinity axiom.

$$\exists x.(\exists y. y \in x \wedge \forall z. z \notin y) \wedge \forall y. y \in x \rightarrow \exists z. z \in x \wedge \forall w.(w \in z \leftrightarrow w \in y \vee w = y)$$

This is a rather complex axiom.

We shall say very little about type theory here. Actually, we prefer to call the formal systems to be considered, higher order logics. In computer science, type theory is often constructive type theory, based on the propositions-as-types idea. It is also called intuitionistic type theory Martin-Löf (1984) or Martin-Löf type theory (see Backhouse et al., 1989), since it was introduced by Per Martin-Löf in 1972. Both extensional and intensional variants of type theory were proposed (intensionality here means that the judgemental equality is understood as definitional equality). Obviously, constructive type theory is very different from the approach by Russell. Since the HANDBOOK OF MATHEMATICAL LOGIC (Barwise, 1977) also has very little on type theory, it seems appropriate to call it *higher order logic* and to reserve *type theory* for constructive type theory, which is an important topic in computer science.

The reference presentation by Joseph (Shoenfield, 1967, ch. 9) tries to justify Zermelo-Fraenkel set theory by constructing the universe of sets in stages and dismisses all alternatives. (Barwise, 1977, page 324) comments:

> It is, of course, possible that there is a completely different analysis of the notion of a set, and this might lead to quite a different set of axioms. Up to the present, however, there has been no analysis of the notion of a set

essentially different from that given here which leads to a satisfactory set of axioms.

This comment, however, implicitly suggests that the axioms of set theory are not fixed once and for ever, but that other foundations of mathematics, not based on set theory as conceived of by Shoenfield, are possible.

In type theory or higher order logic the key idea is that a domain is given and the laws of logic are stated first for this domain and then successively for other domains derived one by one from the original domain and from one another.

2 Axiomatized Mathematics ... –1900

Axiomatization is basically the old idea of going back to first principles. Until 1900 there is — with Frege (1879) as an exception — usually no logic or calculus directly given. From today's point of view some kind of higher order logic is usually assumed. However, it was Hilbert who in 1917–1918 first introduced first order logic as a distinct logic (cp. Eklund, 1996). First order logic was also considered by Löwenheim (1915) in a more restricted manner and by Skolem in (1920; 1922; 1928). But in much of the work by Zermelo, Fraenkel, von Neumann and also in PRINCIPIA MATHEMATICA by Whitehead and Russell (1910–1913) the distinction between first order logic and higher order logic is not clear. Therefore it is axiomatized mathematics rather than formalized mathematics.

In Hilbert and Ackermann's GRUNDZÜGE DER THEORETISCHEN LOGIK from 1928 the exact logic is perfectly clear and also in the completeness result by Gödel (1930). Note that it has been convincingly argued that there was no Skolem-Gödel first order logic proposal at all (Eklund, 1996). In 1931–1936 the notion of (symbolic) logic matured with the very important "appendix" concerning the definition of computability (Turing machines/formal system). So, before 1936 the idea of a formal system/calculus was not clear, although it can actually be traced back to 1666 (at least) in the work of Leibniz (see Leibniz, 1962). As a result much work used the term "formal" in the sense of form/formalism, for example the work of Boole and De Morgan.

2.1 Elementary Logic (ℓ)

Elementary logic is implicit in Frege (1879), but it is not a foundation of mathematics. In fact it is not a fixed language as it needs some predicates. Constants and functions can be added as well, but predicates are needed to be able to talk about them. Here is the calculus for elementary logic (cp. Barwise, 1977).

$$\text{All tautologies are axioms} \qquad \frac{\phi \to \psi \quad \phi}{\psi}$$

$$\frac{\phi \to \psi(v)}{\phi \to \forall x.\psi(x)} \quad v \text{ not free in } \phi \qquad (\forall x.\phi(x)) \to \phi(t)$$

$$\frac{\psi(v) \to \phi}{(\exists x.\psi(x)) \to \phi} \quad v \text{ not free in } \phi \qquad \phi(t) \to \exists x.\phi(x)$$

We later present Prime-Logic P, Supra-Logic Q and Nominalistic Logic NL that extend elementary logic ℓ to a foundation of mathematics. Note that unlike elementary logic, these logics do not have an arbitrary universe.

2.2 Status 1900

Mathematicians like Cantor, Frege, Peano and Hilbert worked intensively on the foundations of mathematics around 1900. There was a critical mass around 1900 to move from the axiomatization of mathematics to the formalization of mathematics.

While considering Cantor's set theory, Russell discovered in 1901 the contradiction that is now called Russell's paradox. The following year Russell saw that the paradox could also be formulated in Frege's logic. Russell proposed type theory in order to avoid this and other paradoxes. Other ways to avoid the paradoxes were soon proposed, in particular axiomatic set theory by Zermelo[2] and others. Russell's book THE PRINCIPLES OF MATHEMATICS appeared in 1903. In the thorough bibliography by Alonzo Church in 1939, this book is the first publication in the history of logic that is marked as recommended for its expository value and furthermore it is also marked as being of special importance in the opinion of the compiler (no later publication has been marked twice). In the introduction to the second edition of the book, which appeared in 1937, Russell stated:

> The fundamental thesis of the following pages, that mathematics and logic are identical, is one which we have never since seen any reason to modify. (Russell, 1903, v)

But in spite of Russell's assurance, both mathematics and logic had changed in the period 1903-1937 and 100 years later it makes even more sense to view mathematics and logic as separate but connected disciplines.

The foundations of mathematics was called "mathematical philosophy". The term is explained by Russell on the very first page of his popular and much translated INTRODUCTION TO MATHEMATICAL PHILOSOPHY (1919):

---

[2] Cp. the contribution of Peckhaus, p. 11–19.

> Mathematics is a study which, when we start from its most familiar portions, may be pursued in either of two opposite directions. The more familiar direction is constructive, towards gradually increasing complexity: from integers to fractions, real numbers, complex numbers, from addition and multiplication to differentiation and integration, and on to higher mathematics. The other direction, which is less familiar, proceeds, by analyzing, to greater and greater abstractness and logical simplicity; instead of asking what can be defined and deduced from what is assumed to begin with, we ask instead what more general ideas and principles can be found, in terms of which what was our starting-point can be defined or deduced. It is the fact of pursuing this opposite direction that characterizes mathematical philosophy as opposed to ordinary mathematics. But it should be understood that the distinction is one, not in the subject matter, but in the state of mind of the investigator.

The term "mathematical philosophy" is rarely used nowadays and if used at all it can also either indicate the use of mathematics in philosophy, for example in ethics, or it can simply be another way of referring to the so-called philosophy of mathematics. For any field of study it is said that there is a corresponding term "philosophy of ..." (sometimes even the philosophy of philosophy is considered). A quick survey yielded thousands of such terms, but one should be careful that philosophers do not impose special norms here.

Russell was influenced by Kant but later he rejected much of the existing work on the philosophy of mathematics. Already in 1900 he stated in a review in MIND:

> It also illustrates the fact that philosophers subsequent to Kant, in writing on mathematics, have thought it unnecessary to become acquainted with the subjects they were discussing, and have therefore left to painful and often crude efforts of mathematicians every genuine advance in mathematical philosophy. (Russell, 1900, 121)

The philosopher Kant was the first to use the terms "analytic" and "synthetic" to divide propositions into types. Kant introduced the analytic/synthetic distinction in the introduction to the CRITIQUE OF PURE REASON (1781). The laws of logic were considered paradigmatic cases of analytic propositions and hence immune to changes. However Quine argued in TWO DOGMAS OF EMPIRICISM (1951) that the different explanations of analyticity were circular. Since no satisfactory explanation of analyticity had been given, Quine took the unit of empirical significance to be the whole of science:

> Epistemologically [classes] are myths on the same footing with physical objects and gods, neither better nor worse except for differences in the degree to which they expedite our dealings with sense experiences. (Quine, 1951, 45)

But this point of view seems to be very different from that generally adopted in 1900.

3 Formalized Mathematics 1901–1950

There was a critical mass around 1900 to move to the formalization of mathematics. Zermelo's paper (1908) on set theory and Russell's paper 1908 on type theory from the same year are important steps. The volumes of PRINCIPIA MATHEMATICA (Whitehead and Russell, 1910–1913) provided much information and inspiration, although the formalization was still problematic. After 1936 at least, the formalization was done properly and the undecidability of first order logic established. Various formulations of set theory in first order logic were proposed, in particular Bernays's reformulation (1937–1954) of von Neumann's work (von Neumann, 1925, 1928b,a, 1929). Gödel showed the consistency of the axiom of choice in 1938 which removed the issue from set theory (except for its independence, but that is less important in this connection).

Set theory began to be directly based on first order logic. There is a reference to Quine's MATHEMATICAL LOGIC from 1940 in ON ZERMELO'S AND VON NEUMANN'S AXIOMS FOR SET THEORY by Hao Wang (1949). This shows that the idea of using first order logic was not yet completely standard since it had to be mentioned at the start, although as it is only explained briefly, it was probably almost standard.

3.1 Prime-Logic (P)

By Prime-Logic we mean the formal system used by Gödel (1931). This builds on Frege's system and Russell's types, but adds Peano's postulates for the natural numbers using the successor operator ′ (prime). — The terms of type 1 are of the form $a, a', a'', a''', \ldots$ where a is either 0 (zero) or a variable of type 1. The terms of type $n > 1$ are just the variables of type n. If in a basic formula $z(x)$ the type of x is n then the type of z must be $n + 1$. — The axioms are as follows:

$p \vee p \to p$ \qquad $p \to p \vee q$
$p \vee q \to q \vee p$ \qquad $(p \to q) \to (r \vee p \to r \vee q)$
$(\forall x.p) \to p[t/x]$ (assuming no free variable in t becomes bound in p)
$(\forall x.p \vee q) \to (p \vee \forall x.q)$ (assuming x is not free in p)
$\exists z.\forall x.z(x) \leftrightarrow p$ (assuming z is not free in p)
$\forall xy.(\forall z.x(z) \leftrightarrow y(z)) \to x = y$ \quad $\forall x.x' \neq 0$
$\forall xy.x' = y' \to x = y$ \quad $\forall z.((z(0) \wedge \forall x.z(x) \to z(x')) \to \forall x.z(x))$

P is an elegant system, but in contrast to both Supra-Logic Q and Nominalistic Logic NL it has a very direct axiom of infinity in the form of Peano's postulates.

3.2 Status 1950

In year 1950 the following six publications set the scene:

- Alfred Tarski and Wanda Szmielew MUTUAL INTERPRETABILITY OF SOME ESSENTIALLY UNDECIDABLE THEORIES, talk given at the 1950 International Congress of Mathematicians (Tarski and Szmielew, 1952).
- Leon Henkin (1950): COMPLETENESS IN THE THEORY OF TYPES.
- ILse L. Novak (Gál): A CONSTRUCTION FOR MODELS OF CONSISTENT SYSTEMS.
- Alan Turing (1950):COMPUTING MACHINERY AND INTELLIGENCE.
- Claude Shannon (1950): PROGRAMMING A COMPUTER FOR PLAYING CHESS.
- Norbert Wiener (1950) THE HUMAN USE OF HUMAN BEINGS: CYBERNETICS AND SOCIETY.

The first three publications contributed with very important new results on the foundations of mathematics and the last three publications introduced very important new concepts to be taken up in computer science. These results and concepts are briefly explained in the following. Directly or indirectly they are central to the understanding of the development of set theory and logic in the period 1951-2000.

Undecidabiliy of Small Fragments of Set Theory

In 1950 Tarski and Szmielew proved that even very small fragments of set theory are essentially undecidable. The small fragment of set theory which they considered has just three axioms: one axiom asserts the existence of an empty set, another asserts that, given two sets x and y, the union of x and the singleton of y exists, while the last axiom is the extensionality axiom which has been shown to be superfluous by the result of Mancini and Montagna (1994). The fragment is so small that it is easily interpretable in virtually every foundational system based on set theory. This result helped make set theory the standard foundation of mathematics.

Completeness of Higher Order Logic

Henkin (1950) proved a completeness result for higher order logic. Church (1940) had presented a new formulation of higher order logic based on the λ-calculus. The resulting logic was in many ways more elegant than Gödel's Prime-Logic P. Despite continued work on higher order logic, for example by Turing (1948), most logicians working on the foundations of mathematics preferred set theory. The completeness result using so-called general models made higher order logic look like a variant of first order logic logic. This result also helped in making set theory the standard foundation of mathematics.

Conservative Extensions

Ilse Novak (Gál) — in connection with work done by Wang and Mostowski — proved that every theorem of (von Neumann-) Bernays-Gödel (NBG) set theory which can be expressed at all in Zermelo-Fraenkel (ZF) set theory is also a theorem of ZF (NBG can be construed as an extension of ZFC *via* the introduction of class variables). NBG is a conservative extension of ZF. This result finally helped make set theory the standard foundation of mathematics, since whether one does set theory in ZF or BG is a matter of taste and one might as well work in the parsimonious ZF (or ZFC).

The Turing Test and Artificial Intelligence

Turing Turing (1950) — using the concept of the universal machine and its ability to imitate any other sort of digital computer — investigated the arguments that had been advanced against the possibility of computers ever being truly intelligent.

The idea of reducing reasoning to computation is an old one — usually called Leibniz's dream — but the computer itself is of course a more recent invention. Babbage described in 1837 the "Analytical Engine" and anticipated virtually every aspect of present-day digital computers but he never built an operational computer. Zuse started construction in 1936 of his first Z-series computer and replaced the hard-to-implement decimal system used by Babbage with the simpler binary system. As a result Zuse's machine was easier to build, although it never worked reliably. One of Zuse's subsequent machines, the Z3, was completed in 1941. It was based on telephone relays and did work satisfactorily. The Z3 thus became the first functional program-controlled, all-purpose, digital computer.

In two 1936 patent applications, Zuse also anticipated that machine instructions could be stored in the same storage used for data. This is the key insight of what became known as the von Neumann architecture and was first implemented in the later IBM Selective Sequence Electronic Calculator (SSEC) in 1948. It was the first computer to run a stored program, although the computer was not fully electronic (just a little ahead of Manchester's Small-Scale Experimental Machine called the 'Baby' — although EDSAC was the world's first practical stored program electronic computer in 1949). The Z4 computer was the world's first commercial digital computer, delivered in 1950. Zuse also claimed to have designed the first higher-level programming language, (Plankalkül), in 1945.

Set theory was also considered by Turing:

> The processes of inference used by the machine need not be such as would satisfy the most exacting logicians. There might for instance be no hierarchy of types. But this need not mean that type fallacies will occur any more than we are bound to fall over unfenced cliffs. Suitable imperatives (expressed within the systems, not forming part of the rules of the system) such as 'Do not use a class unless it is a subclass of one which has been mentioned by teacher' can have a similar effect to 'Do not go too near the edge'. (Turing, 1950, 458)

Turing died in 1954 and therefore he did not attend the Dartmouth College meeting in 1956 that is wrongly thought of as the birthplace of artificial intelligence. Shannon was a key figure at the Dartmouth College meeting and also published a seminal paper in 1950.

Computer Games and Automated Reasoning

In 1950 Shannon published a groundbreaking paper on computer chess and also mentioned automated theorem proving as an area where the proposed techniques were applicable (theorem proving is generally considered to be pioneered by Davis in the early 1950s). Shannon's paper described how a computer could be made to play a reasonable game of chess and is considered to be the first paper on computer chess.[3] Shannon calculated the Shannon number, 10^{120}, as the estimated lower bound on the game-tree complexity of chess:

> The thesis we will develop is that modern general purpose computers can be used to play a tolerably good game of chess. [...] Although perhaps of no practical importance, the question is of theoretical interest, and it is hoped that a satisfactory solution of this problem will act as a wedge in attacking other problems of a similar nature and of greater significance. [...] Chess is generally considered to require "thinking" for skilful play; a solution of this problem will force us either to admit the possibility of a mechanized thinking or to further restrict our concept of "thinking" [...].

Turing (1950) also mentioned chess. The first working AI program, a checkers program written by Strachey, ran on the Manchester Ferranti. In 1952 the program could play a complete game of draughts at a reasonable speed (much later a fully fledged chess program was written by Bernstein for an IBM computer at MIT in 1957). In 1994 the checkers program Chinook was the first computer program to win a human world championship.

Computer Ethics

Wiener (1950) made it clear that the integration of the computer into society will constitute the remaking of society and it is bound to be a multi-faceted, on-going process, which will take decades of effort and will radically change the world. Wiener also developed the field of cybernetics and inspired many to think of computer technology as a means of expanding the capabilities of humans.

---

[3] In 1997, IBM's Deep Blue defeated Kasparov in a six-game match. This was the first time a computer defeated a reigning world champion in a classical chess match. Deep Blue used special chess chips and was immediately dismantled. A few years later the new world champion, Kramnik, was unable to win a chess match against Deep Fritz using a commercial computer (with eight Intel Pentium desktop chips) and with access to the Deep Fritz program for training.

4 Mechanized Mathematics 1951–2000

There was a critical mass around 1950 to move to the mechanization of mathematics. The inventions of the first point-contact transistor in 1947–48 and more importantly the radically different bipolar junction transistor in 1950 — the first type of transistor to be mass-produced and the device which is most commonly referred to as a transistor today — as well as the integrated circuit some ten years later, were critical.

Davis's program for Presburger's algorithm for the theory of addition in first order logic is generally considered as the first mechanization of a formal logic; and the first computer-generated proof confirmed that the sum of two even numbers is again an even number. Automated theorem proving has since been extensively developed and is today based on ingenious search techniques to find a proof for a given theorem in very large search spaces. The resolution principle in particular dominated automated theorem proving after the principle was proposed in Robinson (1965).

Early computers only supported a batch style of working with a long turnaround time. But by the 1960s, a more interactive style was becoming widespread and proof assistants appeared. This idea of a machine and human working together to prove theorems from sketches had already been envisaged by Wang (1960) — *Toward Mechanical Mathematics* in the IBM Journal of Research and Development, 1960 — but this work on automated theorem proving was merely intended to lay the groundwork for such a system (see also McCarthy, 1963). The Automath project of N. G. de Bruijn around 1970 was more modest in its aims with respect to automation but it showed that complete mathematical textbooks could be coded and proof-checked by a computer. First order logic was often used for automated theorem proving but a number of improvements of higher order logic appeared in the period 1951–2000, e.g., Henkin (1963); Andrews (1963, 1972b,a); Gallin (1975); Lambek and Scott (1986).

4.1 Supra-Logic (Q)

By Supra-Logic we mean the formal system proposed by Andrews (1965). This higher order logic is described by Villadsen (2005), where it is used to deal with inconsistent information without rendering the resulting system trivial by the inference *ex falso quodlibet* as in classical logic (paraconsistency). Supra-Logic Q highlights the development from P (Prime-Logic) towards NL (Nominalistic Logic NL), which will be discussed later.

Let α, β, γ, δ, and τ range over types. The transfinite types are built as follows: o is the first finite type (the booleans), σ is the first transfinite type (the supra-booleans), and if α and β are types then $\alpha\beta$ is the type of functions from α to β. The finite types are the types built from o only. $\alpha\beta\ldots\delta\tau$ stands for $\alpha(\beta(\ldots(\delta\tau)))$ and $\langle\alpha\beta\ldots\tau\rangle$ stands for $\alpha\beta\ldots\tau o$. — The operators of propositional logic have the following types: negation $\neg$ has type

oo (or $\langle o \rangle$); conjunction $\wedge$, disjunction $\vee$, implication $\rightarrow$, and biimplication $\leftrightarrow$ all have type ooo (or $\langle oo \rangle$). — Formulas are built from types and terms; a formula is simply a term of type o. A_τ, B_τ, C_τ, D_τ, and E_τ range over terms of type τ. Furthermore A, B, C, D, and E denote the formulas A_o, B_o, C_o, D_o, and E_o, respectively. $A_\alpha B_\beta \ldots D_\delta E_\tau$ stands for $(((A_\alpha B_\beta)\ldots)D_\delta)E_\tau$. — There are primitive variables x, y, z, ... and type variables a, b, c, ... (in the interpretation these range over finite types only). U, V, and W range over type variables and X, Y, and Z range over primitive variables.

The idea behind the transfinite types is that the supra-booleans are the booleans and the boolean functions (hence all the finite types). In order to make this work we introduce a partial order $\geq$ on types which is used in the application term $B_{\gamma\beta} A_\alpha$ to be considered in a moment. — Let the order $>$ on types be the smallest relation such that $\sigma > \alpha$ if α is a finite type; furthermore, $\tau\gamma > \tau\alpha$ if $\gamma > \alpha$ (hence we have that $\sigma > \alpha$ iff α is a finite type). Since there are no types α, β, and γ such that $\alpha > \beta$ and $\beta > \gamma$, the order $>$ is trivially transitive. We write $\alpha \geq \beta$ to indicate that $\alpha > \beta$ or $\alpha = \beta$.

We have equality both as an operator of type $\langle \tau\tau \rangle$ for all types τ and also between types (finite types possibly with type variables). There is λ-abstraction, the choice operator ε and the universal quantifier $\forall$ for types (over finite types only). Hence the terms (and their types) are as follows:

$=_{\langle \tau\tau \rangle}$	$\langle \tau\tau \rangle$	Equality
$\varepsilon_{\langle \tau \rangle \tau}$	$\langle \tau \rangle \tau$	Choice (given a non-empty set, it returns a member)
$B_{\gamma\beta} A_\alpha$	β	Application, $\gamma \geq \alpha$ (for finite types $\gamma \geq \alpha$ iff $\gamma = \alpha$)
X_τ	τ	Called an X-variable
$\lambda X_\tau . A_\alpha$	$\tau\alpha$	Abstraction, no other X-variable free in A_α
$\forall U.A$	o	Universal quantification, no X_τ free in A with U in τ
$\alpha = \beta$	o	Equality between types; α and β finite types

Observe that the formation rules allow X_τ, where X is a primitive variable and τ is a type, possibly with type variables. We omit some obvious types from now on, in particular on bound variables.

The logical operators are introduced as abbreviations:

Nominalistic Logic: From Naive Set Theory to Intensional Type Theory

$A_\alpha \stackrel{\gamma}{=} B_\beta \equiv =_{\langle\gamma\gamma\rangle} A_\alpha B_\beta \quad A_\tau = B_\tau \equiv A_\tau \stackrel{\tau}{=} B_\tau \quad A \leftrightarrow B \equiv A = B$

$\varepsilon X_\tau.A \equiv \varepsilon_{\langle\tau\rangle\tau} \lambda X_\tau.A$

$\top \equiv (\lambda x_o.x) = (\lambda x_o.x) \quad \bot \equiv (\lambda x_o.x) = (\lambda x_o.\top) \quad \neg \equiv \lambda x_o.(x \leftrightarrow \bot)$

$A_\alpha \neq B_\beta \equiv \neg(A_\alpha \stackrel{\gamma}{=} B_\beta) \quad A_\tau \neq B_\tau \equiv \neg(A_\tau = B_\tau) \quad \alpha \neq \beta \equiv \neg(\alpha = \beta)$

$\forall X_\tau.A \equiv (\lambda X_\tau.A) = (\lambda X_\tau.\top) \quad \exists X_\tau.A \equiv \neg \forall X_\tau.\neg A$

$\exists U.A \equiv \neg \forall U.\neg A$

$\wedge \equiv \lambda x_o y_o.(\lambda p_{\langle oo\rangle}.pxy) = (\lambda p_{\langle oo\rangle}.p\top\top) \quad A \wedge B \equiv \wedge AB$

$\rightarrow \equiv \lambda x_o y_o.(x \wedge y \leftrightarrow x) \quad A \rightarrow B \equiv \rightarrow AB$

$\vee \equiv \lambda x_o y_o.\neg(\neg x \wedge \neg y) \quad A \vee B \equiv \vee AB$

The variable binding operators λ, $\forall$, and $\exists$ take a sequence of variables (a sequence would not make sense for ε). The variable binding ε-operator chooses *some* X_τ for which the formula A is satisfied (X_τ can be free in A); if no such X_τ exists, then an arbitrary value is chosen (of the right type). The choice is global in the sense that all choices are the same for equivalent formulas.

As an illustration we consider the following example. Terms interpretable as the number 0 and the successor function S can be defined in various ways in Q, and number theory, with the recursive functions, can be derived. We use the empty n-ary relations as natural numbers as follows. We first introduce a shift operator that takes a term A_α of any finite type α into the transfinite type σ:

$$`A_\alpha\text{'} \equiv \varepsilon x_\sigma . x \stackrel{\sigma}{=} A_\alpha$$

We define the number 0 and the successor function S of types σ and $\sigma\sigma$, respectively.

$$0 \equiv `\bot\text{'} \quad S \equiv \lambda m_\sigma.\varepsilon n_\sigma.\exists a.\exists x_a. m = x_a \wedge n = \lambda y_o.x_a$$

The classical properties of the logical operators can be derived from the following rule and axioms.

$$\frac{C[A_\alpha] \quad A_\alpha \stackrel{\gamma}{=} B_\beta}{C[B_\beta]} \text{ Leibniz's Law}$$

$p_{\langle o\rangle}\top \wedge p_{\langle o\rangle}\bot \leftrightarrow \forall x_o.p_{\langle o\rangle}x \qquad x_\alpha \stackrel{\gamma}{=} y_\alpha \rightarrow (p_{\langle\alpha\rangle}x_\alpha \leftrightarrow p_{\langle\alpha\rangle}y_\alpha)$

$f_{\alpha\beta} = g_{\alpha\beta} \leftrightarrow \forall x_\alpha.f_{\alpha\beta}x = g_{\alpha\beta}x \qquad (\lambda X_\gamma.B_\beta)A_\alpha = B_\beta[A_\alpha/X_\gamma] \quad \boxed{0}$

$p_{\langle\tau\rangle}x_\tau \rightarrow p_{\langle\tau\rangle}(\varepsilon_{\langle\tau\rangle\tau}p_{\langle\tau\rangle}) \qquad \forall U.\top$

$(\forall U.A) \rightarrow A[\tau/U] \quad \boxed{1} \qquad \alpha = \beta \rightarrow (A[\alpha/U] \leftrightarrow A[\beta/U]) \quad \boxed{2}$

$ab = vw \leftrightarrow a = v \wedge b = w \quad \infty \qquad x_a \stackrel{\sigma}{=} y_b \rightarrow a = b$

$o \neq ab \qquad (\forall a.\forall x_a.p_{\langle\sigma\rangle}x_a) \rightarrow \forall y_\sigma.p_{\langle\sigma\rangle}y$

$A[o/U] \wedge (\forall VW.A[V/U] \wedge A[W/U] \rightarrow A[VW/U]) \rightarrow \forall U.A \quad \boxed{3}$

$\boxed{0}$ Free occurrences of X_γ in B_β must be free for free variables in A_α
$\boxed{1}$ Free occurrences of U in A must be free for variables in τ
$\boxed{2}$ Free occurrences of U in A must be free for variables in α and β
$\boxed{3}$ Free occurrences of U in A must be free for distinct variables V and W

Of course the axiom ∞ is a kind of axiom of infinity, but since it operates on the finite types only it can be regarded as "logical" with respect to the type logic. Given the usual properties of equality and λ-abstractions it is quite straightforward to see that we obtain a classical logic from the abbreviations above.

4.2 Status 2000

Almost all of the seventeen provers for mathematics use axiomatic set theory or higher order logic (Wiedijk 2006; some use first order logic but are not able to express statements about, say, the square root function). Furthermore, six of the seven provers with a large mathematical standard library use higher order logic — the only exception with respect to this is Mizar (Matuszewski and Rudnicki, 2005).

5 Integrated Mathematics 2001–...

There was a critical mass around 2000 to move to the integration of mathematics. Instead of stand-alone systems for the mechanization of mathematics, the aim for the future is to make the knowledge available in a new, integrated way, taking advantage of the internet and the semantic web. The combination of mathematics and computer science is often in focus. So, in 2001 for instance, the Mathematical Knowledge Management (MKM) conferences started. The so-called QED Manifesto also played a major role (Wiedijk, 2007).

5.1 The Prospects for the Future

At the annual meeting of the Association of Symbolic Logic in 2000 there was a panel discussion on "The Prospects for Mathematical Logic in the Twenty-First Century" where Sam Buss stated:

> Indeed, it is likely that there are more people working on logic within computer science than outside of computer science. (Buss et al., 2001, 14)

And Buss continued with three predictions:

- Prediction 1 was that the P *versus* NP problem would be solved in 2010 (+/−10). Preferably it would involve some kind of diagonalization and thereby boost the importance of logic. Buss believed that P and NP are distinct and that related questions in complexity theory would be solved around the same time; obviously very important in computer science.

- Prediction 2 was that there would be limited but significant success in artificial intelligence by 2050 (+/−30). Buss believed that logic-based reasoning would be needed.
- Prediction 3 was that computer databases of mathematics would be possible in 2030 (+/−10). Such databases should provide an integration of mathematics.

Notice that predictions 2 and 3 are independent.

5.2 Nominalistic Logic (NL)

We have considered the historical development which started in 1901 with naive set theory and leads up to Gilmore's 2001 presentation of Intensional Type Theory (ITT) in 2001, which will be discussed now. ITT yields a logical foundation of mathematics without a direct axiom of infinity, and this is one of the advantages of ITT.

ITT conceives applied logic – and "ordinary" mathematics – as dialects of natural language and offers an explanation for the paradoxes of set theory as a confusion of use and mention. The close connection between mathematics and language is supported by the book *Math Gene* by Devlin (2000), whose main claim is that the feature of the brain that enables language is the same feature that makes it possible to do mathematics. In both cases the ability to formulate and follow complicated plans — with iterations and alternatives — is essential. In other words, the notion of computation is the key. The close connection between mathematics and language supports the move to integration of mathematics and to a simple and elegant logic as the meeting point. Russell is well known for the claim that mathematics and logic are identical. The claim, in one form or another, is central to logicism as it came to be called by Abraham Fraenkel in 1928 and later by Rudolf Carnap (see Grattan-Guinness 2000, pages 479 and 501 and Gilmore 2005).

As the name suggests, ITT emphasizes intensionality and has so-called intensional rules, but we find that many of these features are better understood as nominalization. A nominal is a word or phrase that functions as a noun. By nominalization we understand the process of converting a word or phrase into a noun. In many logics with types for individuals and predicates, nominalization can be seen as using predicates as individuals. Gilmore's nominalistic conceptions goes back to 1950–1951 (Gilmore, 2005, page ivx).

Nominalistic Logic is a new presentation of ITT as a sequent calculus. In most formulations of finite type theory, in particular logics based on simple type theory (cf., *e.g.*, Church, 1940), nominalization is not possible because each term has a single type only. In transfinite type theory, for example the formulation given by Andrews (1965) which makes use of type variables, all finite types are included in the first transfinite type, but no nominalization is possible for the transfinite types as such. However, the inclusion just mentioned yields a logical foundation of mathematics without a direct axiom of infinity.

With respect to mathematics, nominalism is taken to be the view that there are no abstract objects like numbers and sets. More precisely such abstract objects have no independent existence but exist only as names. A nominalist can still talk about abstract objects. This is similar to the situation where one talks about unicorns, even though one does not believe that they exist. Nominalism in general can be seen in contrast to realism which holds that when we use descriptive words such as "red" for certain objects, it is because all the red things are red in virtue of the existence of a universal; a single abstract object in this case that is a part of all the red things. The distinction between realism and nominalism has a long history. The medieval philosopher William of Ockham was a pioneer of nominalism. Ockham's Razor, the idea of parsimony in explanation and theory building, is important to modern science. Since abstract objects lack causal powers, nominalists think that it is appropriate to use Ockham's Razor and simply get rid of them.

In computer science we find the distinction between concrete and abstract objects problematic. Is a star a concrete object? An atom? Since computers are consummate nominalists, nominalistic interpretations of languages intended for computer applications are needed. For it can be argued that the concept of an abstract object is so unclear that there is no objective, agreed condition that would need to be satisfied in order for the contention that there are abstract objects to be true. Recall in this connection Russell's famous saying in RECENT WORK ON THE PRINCIPLES OF MATHEMATICS from 1901:

> Thus mathematics may be defined as the subject where we never know what we are talking about, nor whether what we are saying is true.

The purpose here is to define the logic NL consisting of a language and a calculus. A concise "stand-alone" definition suitable for the integration of mathematics is provided. It consists of five figures:

- Fig. 1 defines the terms and the concept of λ-conversion.
- Fig. 2 defines the types and the concept of a sentence.
- Fig. 3 defines some basic abbreviations in order to get the usual boolean operators and quantifiers.
- Fig. 4 defines a calculus of sequents $\Gamma \vdash \Delta$ for formula sequences Γ and Δ and gives a proof system for sequents of the special form $\vdash \theta$, where θ is a sentence.
- Fig. 5 defines some additional abbreviations in order to get the number zero and the successor function for numbers.

Furthermore a collection of models as well as a corresponding relation $\models$ of satisfaction can be defined such that $\models \theta$ means that θ holds in all models. Soundness and completeness — thus $\vdash \theta$ iff $\models \theta$ — is provable by standard means (cf. Gilmore, 2001, 2005).

Nominalistic Logic: From Naive Set Theory to Intensional Type Theory

---

c and $x, y, z, \ldots, x_n, y_n, z_n, \ldots$ range over constants and countably infinitely many variables, respectively, and $p, q, r, s, t, \ldots, p_n, q_n, r_n, s_n, t_n, \ldots$ range over terms produced by the grammar:

$$t ::= pt \mid \lambda x.p \mid x \mid c$$

$pt_1 \cdots t_n$ stands for $(pt_1) \cdots t_n$

$\lambda x_1 \cdots x_n.p$ stands for $\lambda x_1. \cdots \lambda x_n.p$ (used for all variable-binding operators).

A variable x occurs bound (free) in a term if x is (not) in the scope of a λx. The term $p[t/x]$ is p with every free occurrence of x replaced with t.

α-renaming of the variable x to the variable y in $\lambda x.p$ is $\lambda y.p[y/x]$ assuming y is not free in p and does not become bound in p.

β-reduction of $(\lambda x.p)t$ is $p[t/x]$ assuming no free variable in t becomes bound.

η-reduction of $\lambda x.px$ is p assuming x is not free in p.

$s \sim t$ if t is s with a series of α-renamings.

$s \succ t$ if t is s with either a β-reduction or an η-reduction.

Fig. 1. NL terms

NL terms

See Fig. 1 for the details. We start by clarifying the use of meta-variables. As regards constants, a single meta-variable c ranging over them suffices. Countably infinitely many constants, all of different types, are introduced later and used in such a way that in each situation the type determines the constant. Furthermore, the use of operators further eliminates the use of constants. With respect to variables, however, countably infinitely many meta-variables x, y, z, ... ranging over them are introduced. For simplicity these meta-variables are also used as ordinary variables where appropriate, that is, in abbreviations like $\top := \exists x.x$, and in results like $\vdash \forall x.\exists y.x \neq y$ (if no meta-variables were used, then it would be necessary to require that x and y were different variables, since $\vdash \forall x.\exists x.x \neq x$ would be inappropriate).

Note that in grammar rules like $t ::= pt \mid \ldots$ all meta-variables are independent and hence $t ::= tt \mid \ldots$ would make no difference except that it would not indicate that the first term (viz. p in the original rule) must have a predicate type (this is enforced by the type system to be introduced later). Thus compare the definition given in Fig. 1 with the corresponding definition for the pure untyped λ-calculus:

$$e ::= ee \mid \lambda v.e \mid v$$

$\tau, \ldots, \tau_n, \ldots$ and $\sigma, \ldots, \sigma_n, \ldots$ range over types and predicate types, respectively, produced by the grammar:

$$\tau ::= \sigma \mid \imath$$

$$\sigma ::= \tau\sigma \mid o$$

$\tau_1 \cdots \tau_n \sigma$ stands for $\tau_1 \cdots (\tau_n \sigma)$

It is assumed that every constant and variable has a unique type such that there are infinitely many variables for each type

The type system $t : \tau$ determines that a term t has type τ by the conditions:

$x : \tau$ if x has type τ	(var)
$c : \tau$ if c has type τ	(con)
$p : \tau\sigma$ and $t : \tau$ gives $pt : \sigma$	(app)
$x : \tau$ and $p : \sigma$ gives $\lambda x.p : \tau\sigma$	(abs)
$p : \imath$ if p is nominalizable	(nom)

p is nominalizable if $p : \sigma$ and $x : \imath$ for every free variable x in p

p is a formula if $p : o$ (a sentence is a closed formula)

Fig. 2. NL types

A term t is in normal form if there are no terms t_1 and t_2 such that $t \sim t_1$ and $t_1 \succ t_2$. Note that $t \sim t$ but not $t \succ t$.

NL types

See Fig. 2 for details. The type system has rules for variables (var), constants (con), applications (app), abstractions (abs), and nominalizations (nom). — Compare this system with the definition of the higher-order boolean types:

$$\rho ::= \rho\rho \mid o$$

(Special) predicates of type τ have the "dual type" (see Gilmore, 2001, 386f) $\imath$ for names of predicates as individuals *via* nominalization, but is allowed for arguments only.

Nominalistic Logic: From Naive Set Theory to Intensional Type Theory

There is a constant of type ooo and for every type τ a constant of type $(\tau o)o$

The constants and variables must have appropriate types in the abbreviations:

$$\neg p := cpp \qquad \text{Here } c \text{ means "neither } \ldots \text{ nor } \ldots\text{"}.$$
$$p \vee q := \neg cpq \qquad \text{Ditto.}$$
$$p \wedge q := \neg(\neg p \vee \neg q)$$
$$\exists x.p := \neg c\lambda x.p \qquad \text{Here } c \text{ means "no } \ldots \text{ exists"}.$$
$$\forall x.p := \neg \exists x.\neg p$$
$$s \neq t := (\lambda xy.\exists z.zx \wedge \neg zy)st$$
$$s = t := \neg(s \neq t)$$
$$p \to q := \neg p \vee q$$
$$p \leftrightarrow q := (p \to q) \wedge (q \to p)$$

The operator priority is as follows from high to low: $\neg = \wedge \vee \to \leftrightarrow \lambda$. Variable-binding operators, like $\exists$ and $\forall$, have the same priority as λ. Other operators, like $\neq$, have the same priority as $=$ (or $\neg$ if unary).

$\vee$ and $\wedge$ are left associative and $\to$ and $\leftrightarrow$ are right associative.

$\doteq$ and $\neq$ are similar to $=$ and $\neq$, respectively, but always has type uo.

Fig. 3. NL basic abbreviations

NL basic abbreviations

See Fig. 3 for details. Recall that a single meta-variable c is used for all constants (due to the types, no confusion can arise).

Instead of equality we find non-equality easier to grasp due to its "constructive" character: two objects are different if and only if one can come up with a property where they differ (cf. Leibniz's principle). We therefore start with non-equality $\neq$ and take equality $=$ to be the negation hereof (we use the usual symbols even though we consider non-equality as more basic).

NL calculus

See Fig. 4 for details. The sequent calculus for NT is a modification of existing sound and complete systems for ITT (Gilmore, 2001, 2005). — We conceive a sequent $\Gamma \vdash \Delta$ as a complex of two sequences of alternatives Γ, Δ where for any interpretation (at least) one of the formulas in Δ must be true or one

Θ, Γ and Δ range over possible empty formula sequences

The sequent calculus $\Gamma \vdash \Delta$ has sequences on both sides in the rules:

$$p \vdash q \quad \text{if } p \sim q \text{ or } p \succ q \tag{S}$$

$$\frac{\Gamma \vdash \Delta}{p, \Gamma \vdash \Delta \quad \Gamma \vdash \Delta, p} \tag{T}$$

$$\frac{\Theta, p, q, \Gamma \vdash \Delta}{\Theta, q, p, \Gamma \vdash \Delta} \quad \frac{\Gamma \vdash \Delta, p, q, \Theta}{\Gamma \vdash \Delta, q, p, \Theta} \tag{E}$$

$$\frac{p, p, \Gamma \vdash \Delta}{p, \Gamma \vdash \Delta} \quad \frac{\Gamma \vdash \Delta, p, p}{\Gamma \vdash \Delta, p} \tag{C}$$

$$\frac{\Gamma \vdash \Delta, p, q}{cpq, \Gamma \vdash \Delta} \quad \vdash p, q, cpq \tag{P}$$

$$cp, pt \vdash \qquad \frac{px, \Gamma \vdash \Delta}{\Gamma \vdash \Delta, cp} \quad x \text{ not free in } p, \Gamma \text{ or } \Delta \tag{Q}$$

$$\vdash p = q \leftrightarrow p \doteq q \tag{N}$$

The left and right rules are placed next to each other.

Double lines indicate that the rule works in both directions.

Multiple conclusions indicate multiple rules each with a single conclusion.

Fig. 4. NL calculus

of the formulas in Γ must not be true (hence false). The order and possible repetitions of formulas in the sequences do not matter.

The rules are as follows:

- Start (S).
- Thin (T) — also the cut rule in opposite direction.
- Exchange (E).
- Contract (C).
- Proposition (P).
- Quantification (Q).

$$\begin{aligned}
\top &:= \exists x.x & &\text{Truth}\\
\bot &:= \neg\top & &\text{Falsity}\\
s \equiv t &:= (\lambda xy.\forall z_1\cdots z_n.xz_1\cdots z_n \leftrightarrow yz_1\cdots z_n)st & &\text{Equivalence}\\
s \not\equiv t &:= \neg(s \equiv t)\\
t' &:= (\lambda xy.x \doteq y)t & &\text{Successor}\\
0 &:= \lambda x.x \neq x & &\text{Zero}\\
1 &:= 0'\\
2 &:= 1'\\
3 &:= 2' \quad \cdots
\end{aligned}$$

Fig. 5. NL additional abbreviations

- Nominalization (N).

We omit the axiom of choice for simplicity; see Gilmore (2005) for more on choice terms.

The left (P) rule can be replaced with the rules $cpq, p \vdash$ and $cpq, q \vdash$, and the right (P) rule with the following rule:

$$\frac{p, \Gamma \vdash \Delta \quad q, \Gamma \vdash \Delta}{\Gamma \vdash \Delta, cpq}$$

Similarly for the left (Q) rule:

$$\frac{\Gamma \vdash \Delta, pt}{cp, \Gamma \vdash \Delta}$$

But the right (Q) rule cannot be replaced in a similar manner to the left (P) rule.

The rule (N) ensures that nominalizable terms are equal if and only if they are equal as predicates.

NL additional abbreviations

See Fig. 5 for the details. Some results:

$$\vdash 0 = 0$$
$$\vdash 0 \neq 1$$

But not $\vdash 0 = 1$ (unless ZFC, the standard foundation of mathematics, is inconsistent, but this possibility is ignored here).

$$\vdash \forall x.x' \neq 0$$

$$\vdash \forall xy.x' = y' \to x = y$$

The last result can also be stated as the contraposition:

$$\vdash \forall xy.x \neq y \to x' \neq y'$$

Of course we also have the result:

$$\vdash \forall xy.x = y \to x' = y'$$

Besides equality, there is the usual extensional equality or equivalence $\equiv$. Extensional equality is only possible for predicates, whereas (intensional, in order to distinguish) equality is possible for all terms, including individuals. Extensionality can be characterized in a form that says something directly about extensions of predicates:

$$\forall xy.x \neq y \to x \not\equiv y$$

In other words (intensional) non-equality implies extensional non-equality. For many applications, in particular in computer science and artificial intelligence, extensionality is problematic. For example, we take the predicates "professor" and "male" as (intensionally) non-equal, but in a particular situation, say with respect to a given university database, the predicates could be extensionally equal. We simply take intensionality to be the lack of extensionality. Since it is not possible to point directly at something as establishing intensionality, rules of intensionality (Gilmore, 2001, sec. 1) do not seem to make sense. But in NL neither $\vdash \forall xy.x \neq y \to x \not\equiv y$ nor $\vdash \forall xy.x \equiv y \to x = y$ (again assuming the consistency of ZFC).

6 Conclusion

Our aim has been to present a succinct view of the development of mathematics, logic and set theory from the discovery of Russell's paradox in 1901 to recent work on type theory, in particular Nominalistic Logic (NL) based on Intensional Type Theory (ITT) as put forward by Gilmore (2001). The modern discipline of logic was developed in the period between 1879 and 1931, with the seminal publications by Frege and Gödel as natural reference points (cf. van Heijenoort, 1967). Logic is the study of formal reasoning systems, or simply logics, like first order logics and higher order logics.

Unlike Prime-Logic P but like Supra-Logic Q, based on transfinite type theory (Andrews, 1965), NL yields a logical foundation of mathematics without a direct axiom of infinity. But the transfinite types add extra complexity. However, the nominalistic interpretation of predication appears to have many applications, for instance in programming language semantics, advanced databases and natural language understanding Gilmore (2005).

Finally, we see a possibility to combine the (paraconsistent) higher order logic described in Villadsen (2004) with NL. This would mean that the basic types ι and o unify (classical logic is retained as a special case). In the end, we agree with the comment made by Andrews (2002) in the preface of AN INTRODUCTION TO MATHEMATICAL LOGIC AND TYPE THEORY Andrews (2002):

> One of the basic tasks of mathematical logic is the formalization of mathematical reasoning. Both type theory [...] and [...] set theory can be used as formal languages for this purpose, and it is really an accident of intellectual history that at present most logicians and mathematicians are more familiar with axiomatic set theory than with type theory.

Acknowledgements

This article is partly based on the invited presentation ON INTENSIONAL TYPE THEORY at the International Conference on Computational, Foundational and Philosophical Issues in Non-Standard Set Theories, 2 June 2006, Kolding, Denmark, and the subsequent short presentation NOMINALIZATION IN INTENSIONAL TYPE THEORY at the Twenty-First Annual IEEE Symposium on Logic in Computer Science (LICS), 12–15 August 2006, Seattle, USA.

Thanks to Paul Gilmore, Klaus Grue and Reinhard Muskens for useful discussions and in particular also thanks to Klaus Robering for support in many ways. Finally thanks to the anonymous reviewer for comments.

References

Peter Bruce Andrews. A reduction of the axioms for the theory of propositional types. *Fundamenta mathematicae*, 52:345–350, 1963.

Peter Bruce Andrews. *A transfinite type theory with type variables*. North-Holland, Amsterdam, 1965.

Peter Bruce Andrews. General models and extensionality. *The journal of symbolic Logic*, 37(2):395–397, 1972a.

Peter Bruce Andrews. General models, descriptions, and choice in type theory. *The journal of symbolic Logic*, 37(2):385–394, 1972b.

Peter Bruce Andrews. *An introduction to mathematical logic and type theory: to truth through proof*. Kluwer Academic Publishers, Dordrecht, 2nd edition, 2002.

Roland Carl Backhouse, Paul Chisholm, Grant Malcolm, and Erik Saaman. Do-it-yourself type theory. *Formal aspects of computing*, 1:19–84, 1989.

Jon Barwise, editor. *Handbook of mathematical logic*. North-Holland, Amsterdam, 1977.

Paul Bernays. A system of axiomatic set theory. *JSL*, 2, 6, 7, 8, 13, 19: 65–77(2), 1–17(6), 65–89(7), 133–145(7), 89–106(8), 65–79(13), 81–96(19), 1937–1954.

Samuel R. Buss, Alexander S. Kechris, Anand Pillay, and Richard A. Shore. The prospect for mathematical logic in the twentyfirst century. *The Bulletin of Symbolic Logic*, 7:169–196, 2001.

Alonzo Church. *A bibliography of symbolic logic (1666–1935)*. Association for symbolic logic, Providence RI, 1939. Originally published in several parts from 1936–1938 in *The journal of symbolic logic*.

Alonzo Church. A formulation of the simple theory of types. *The journal of symbolic Logic*, 5:56–68, 1940.

Paul J. Cohen. The independence of the continuum hypothesis, i. *Proceedings of the National Academy of Sciences of the U.S.A*, 50:1143–1148, 1963.

Keith Devlin. *The math gene*. Basic Books, New York, 2000.

Matti Eklund. On how logic became first-order. *Nordic journal of philosophical logic*, 1(2):147–167, 1996.

Gottlob Frege. *Begriffsschrift, eine der arithmetischen nachgebildete Formelsprache des reinen Denkens*. Nebert, Halle a.d.S., 1879. English translation van Heijenoort (1967, 5–82).

Gottlob Frege. *Grundgesetze der Arithmetik, begriffsschriftlich abgeleitet, 2 Bde*. Pohle, Jena, 1893 and 1903. English translation of selected parts. *Basic laws of arithmetic*. University of California Press, Berkeley and Los Angeles 1967.

Daniel Gallin. *Intensional and higher-order modal logic*. North-Holland, Amsterdam, 1975.

Paul Carl Gilmore. An intensional type theory: motivation and cut-elimination. *The journal of symbolic Logic*, 66(1):383–400, 2001.

Paul Carl Gilmore. *Logicism renewed: logical foundations for mathematics and computer science*. A. K. Peters, La Jolla, CA, 2005.

Kurt Gödel. Die Vollständigkeit der Axiome des logischen Funktionenkalküls. *Monatshefte für Mathematik und Physik*, 37:349–360, 1930. Reprinted with parallel English translation in Gödel (1986, 102–123).

Kurt Gödel. Über formal uentscheidbare Sätze der *Principia mathematica* und verwandter Systeme I. *Monatshefte für Mathematik und Physik*, 38: 173–198, 1931. Reprinted with parallel English translation Gödel (1986, 144-195).

Kurt Gödel. The consistency of the axiom of choice and of the generalized continuum hypothesis. *Proceedings of the National Academy of Sciences of the U.S.A*, 24:556–557, 1938. Reprinted Gödel (1990, 26–27).

Kurt Gödel. *Collected works, vol. I. Publications 1929–1936*. Oxford University Press, New York and Oxford, 1986.

Kurt Gödel. *Collected works, vol. II. Publications 1938–1974*. Oxford University Press, New York and Oxford, 1990.

Ivor Grattan-Guinness. *The search for mathematical roots, 1870–1940: logics, set theories and the foundations of mathematics from Cantor through

Russell to Gödel. Princeton University Press, Princeton NJ and Oxford, 2000.

Leon Henkin. Completeness in the theory of types. *The journal of symbolic Logic*, 15:87–110, 1950.

Leon Henkin. A theory of propositional types. *Fundamenta mathematicae*, 52:323–344, 1963.

David Hilbert and Wilhelm Ackermann. *Grundzüge der theoretischen Logik*. Springer, Berlin, 1928.

Immanuel Kant. *Kritik der reinen Vernunft*. Hartknoch, Riga, 1781. English translation by Norman Kemp Smith. *Critique of pure reason*. 2nd edition. Palgrave Macmillan, Basingstoke 2007.

Joachim Lambek and Philip J. Scott. *Introduction to higher order categorical logic*, volume 7. Cambridge University Press, Cambridge GB, 1986.

Gottfried Wilhelm Leibniz. *Dissertatio de arte combinatoria (1666)*. Olms, Hildesheim and New York, 1962. *Mathematische Schriften*. Ed. by C. I. Gerhardt. Vol. 5:8–79. Originally this collection appeared in Halle 1858.

Leopold Löwenheim. Über Möglichkeiten im Relativkalkül. *Mathematische Annalen*, 76:447–470, 1915. English translation van Heijenoort (1967, 232–251).

Antonella Mancini and Franco Montagna. A minimal predicative set theory. *Notre Dame journal of formal logic*, 35(2):186–203, 1994.

Per Martin-Löf. *Intuitionistic type theory*. Bibliopolis, Naples, 1984.

Roman Matuszewski and Piotr Rudnicki. Mizar: the first 30 years. *Mechanized mathematics and its applications*, 4:3–24, 2005.

McCarthy. Towards a mathematical science of computation. In Paul Braffort and David Hirshbor, editors, *Computer programming and formal systems*, pages 33–70. NH, Amsterdam, 1963. Previous versions of this article had been presentend in 1961 and 1962.

Ilse L. Novak (Gál). A construction for models of consistent systems. *Fundamenta mathematicae*, 37:87–110, 1950.

Willard Van Orman Quine. *Mathematical logic*. Harvard University Press, Cambridge MA, 1940. 2nd, revised edition 1951.

Willard Van Orman Quine. Two dogmas of empiricism. *Philosophical Review*, 60:20–43, 1951. Cited according to the reprint in Quine's *From a logical point of view*. Harvard University Press, Cambridge MA 1953, 20–46.

John Alan Robinson. A machine-oriented logic based on the resolution principle. *Journal of the ACM*, 12(1):23–41, 1965.

John Barkley Rosser and Hao Wang. Non-standard models of formal logics. *JSL*, 15:113–129, 1950.

Bertrand Russell. Review of Julius Schulz *Psychologie der Axiome*. Vandenhoeck & Rupprecht, Göttingen, 1899. *Mind*, 9:118–121, 1900.

Bertrand Russell. *The principles of mathematics*. Allen & Unwin, London, 1903. 2nd edition 1937.

Bertrand Russell. Mathematical logic as based on the theory of types. *American journal of mathematics*, 30:222–262, 1908.

Bertrand Russell. *Introduction to mathematical philosophy*. Allen & Unwin, London, 1919.

Claude Shannon. Programming a computer for playing chess. *Philosophical Magazine, Ser. 7*, 41:256–275, 1950.

Joseph R. Shoenfield. *Mathematical logic*. Addison-Wesley, Reading MA, 1967.

Thoralf Skolem. Logisch-kombinatorische Untersuchungen über die Erfüllbarkeit oder Beweisbarkeit mathematischer Sätze nebst einem Theoreme über dichte Mengen. *Videnskapsselskapets skrifter, I. Matematisknaturvidenskabelig klasse*, no. 3, 1920. English translation of the first section van Heijenoort (1967, 252–263).

Thoralf Skolem. Einige Bemerkungen zur axiomatischen Begründung der Mengenlehre. In *Matematikerkongressen i Helsingfors den 4–7 Juli 1922. Den femte skandinaviska matematikerkongressen, Redorgörelse*, pages 217–232, Helsinki, 1922. Akademiska Bokhandlen. English translation van Heijenoort (1967, 290–301).

Thoralf Skolem. Über die mathematische Logik. *Norsk matematisk tidsskrift*, 10:125–142, 1928. English translation van Heijenoort (1967, 508–504).

Alfred Tarski and Wanda Szmielew. Mutual interpreatibiltiy of some essential undecidabel theories. In *Proceedings of the International Congress of Mathematicians, Cambridge MA, August 30–September 6, 1950*, pages 705–720, Providence RI, 1952. American Mathematical Society.

Alan Turing. Practical forms of type theory. *The journal of symbolic Logic*, 13:80–94, 1948.

Alan Turing. Computing machinery and intelligence. *Mind*, 59:433–460, 1950.

Jean van Heijenoort, editor. *From Frege to Gödel — a source book in mathematical logic (1879–1931)*. Harvard University Press, Cambridge MA, 1967.

Jørgen Villadsen. A paraconsistent higher order logic. In Bruno Buchberger and John A. Campbell, editors, *Artificial intelligence and symbolic computation*, pages 38–51, Berlin and Heidelberg and New York, 2004. Springer.

Jørgen Villadsen. Supra-logic: using transfinite type theory with type variables for paraconsistency. *Journal of applied non-classical logics*, 15(1):45–58, 2005.

John von Neumann. Eine Axiomatisierung der Mengenlehre. *Jorunal für die reine und angewandte Mathematik*, 154:219–240, 1925. English translation (van Heijenoort, 1967, 34–56).

John von Neumann. Die Axiomatisierung der Mengenlehre. *Mathematische Zeitschrift*, 27:669–752, 1928a.

John von Neumann. Über die Definition durch transfinite Induktion und verwandte Fragen der allgemeinen Mengenlehre. *Mathematische Annalen*, 99:373–391, 1928b.

John von Neumann. Über eine Widerspruchsfreiheitsfrage in der axiomatischen Mengenlehre. *Journal für die reine und angewandte Mathematik*, 160: 227–241, 1929.

Hao Wang. On Zermelo's and von Neumann's axioms for set theory. *Proceeding of the National Academy of Sciences of the U.S.A*, 35:150–155, 1949.

Hao Wang. The formalization of mathematics. *The journal of symbolic logic*, 19:241–266, 1954.

Hao Wang. Towards mechanical mathematics. *IBM Journal of Research and Development*, 4:2–22, 1960.

Hermann Weyl. Review of P. A. Schilpp *The philosophy of Bertrand Russell*. Library of living philosophers, vol. 5. Northwestern University Press 1944. *American Mathematical Monthly*, 53:210, 1946.

Alfred Whitehead and Bertrand Russell. *Principia mathematica, vol. 1–3*. Cambridge University Press, Cambridge GB, 1910–1913.

Freek Wiedijk, editor. *The Seventeen Provers of the World*. Springer, Berlin and Heidelberg and New York, 2006. LNCS 3600.

Freek Wiedijk. The QED Manifesto revisited. *Studies in Logic, Grammar and Rhetoric*, 10(23):121–133, 2007.

Norbert Wiener. *The human use of human beings: cybernetics and society*. Da Capo Press, New York, 1950.

Ernst Zermelo. Untersuchungen über die Grundlagen der Mengenlehre, I. *Mathematische Annalen*, 65:261–281, 1908.

A Gentle Introduction to Map Theory

Klaus Grue

Department of Computer Science
University of Copenhagen

grue@diku.dk

Summary. Map theory (MT) is a foundation of mathematics. MT is slightly more powerful than ZFC set theory, but its merit is that it builds on top of computable functions where ZFC builds on top of finite sets. Thus, where ZFC is suited as a foundation of mathematics, MT is suited as a foundation of both mathematics and computer science. As a foundation of computer science, MT is unusual in that it allows classical reasoning rather than being restricted to constructive reasoning.

This paper gives an introduction to MT intended for a broad audience consisting of mathematicians, logicians, and computer scientists. For the sake of readers who know ZFC already, the introduction given in the present paper is based on ZFC. But this does not change the fact that MT is a foundation in its own right in the sense that one can state and prove all theorems of classical mathematics from the axioms and inference rules of MT.

1 Introduction

1.1 ZFC

Recall that ZFC is a generalization of the theory of finite sets:

- The power set of a finite set is finite. Therefore, ZFC has a power set operator.
- The union of a finite set of finite sets is finite. Therefore, ZFC has a union operator.
- The complement of a finite set is not finite. Therefore, ZFC has no complement operator.

Even the axiom of choice, which is also known as the axiom of Zermelo, is valid in the world of finite sets. The only axiom of ZFC which transcends the world of finite sets is the axiom of infinity.

Map theory is constructed the same way, except that it begins with computable functions.

1.2 Overview

We introduce map theory in the following steps:

Section 2 defines a programming language M for expressing computable functions. The notion of computable functions is much richer than that of finite sets, and for that reason we shall dwell longer on computable functions in MT than it is usual to dwell on finite sets in ZFC.

Section 3 presents a theory of M, *i.e.*, a theory which describes the behavior of programs expressed in the M programming language. Section 3 also defines a relation of behavioral equivalence of programs. Programs *modulo* this equivalence relation are referred to as maps.

Section 4 presents a mental model for visualizing maps. That mental model can be extended into a firm, mathematical model, but we shall not do so in the present paper.

Section 5 discusses which quantifiers can be added to the M programming language. Section 5 also states the four axioms which turn the theory from Section 3 into Map Theory. The axioms do not complete the definition of MT, however, since they refer to a construct ϕ which is defined in Section 6.

Section 6 selects the domain of quantification. It turns out to be a bad idea to let the quantifiers of Map Theory quantify over all maps. Rather, Section 6 defines the notion of a 'wellfounded' map and lets quantifiers quantify over these maps. Section 6 also defines a predicate ϕ for testing whether or not a map is wellfounded.

MT is a foundation which can stand on its own feet. But we utilize the readers previous knowledge and use ZFC as a quick way of explaining MT.

1.3 History

MT was introduced in 1992 in Grue (1992). Grue (1992) proved all axioms and inference rules of ZFC in MT and also proved the consistency of MT in ZFC+SI where SI is the assumption that there exists an inaccessible ordinal.

Berline and Grue (1997) showed that MT has a quite natural Scott model. That made MT much easier to understand.

MT from 1992 has a peculiar, *ad hoc* list of 'construction' axioms. A version where all the construction axioms are replaced by a definition was published in Grue (2002a), but the consistency has never been proved and a model of the version in Grue (2002a) is likely to be much less natural than the one in Berline and Grue (1997).

The present paper proposes a new axiomatization. This new version still replaces the construction axioms by a definition. But the model in Berline and Grue (1997) is expected to satisfy the new version. Furthermore, the new version is expected to be able to prove all axioms and inference rules of ZFC. Verification of the last two claims, however, is future work.

2 The M programming language

2.1 Syntax

The programming language M underlying MT has the following BNF syntax:[1]

$$
\begin{aligned}
&\mathcal{V} ::= \mathsf{v}_1 \mid \mathsf{v}_2 \mid \mathsf{v}_3 \mid \cdots && \text{(object Variable)} \\
&\mathcal{U} ::= \mathsf{T} && \text{(Urelement)} \\
&\mathcal{F} ::= \lambda \mathcal{V}.\mathcal{T} && \text{(Function)} \\
&\mathcal{N} ::= \mathcal{U} \mid \mathcal{F} && \text{(Normal form)} \\
&\mathcal{A} ::= \mathcal{T}\mathcal{T} && \text{(functional Application)} \\
&\mathcal{B} ::= \mathsf{if}(\mathcal{T},\mathcal{T},\mathcal{T}) && \text{(Branch)} \\
&\mathcal{D} ::= \mathcal{T}\|\mathcal{T} && \text{(parallel Disjunction)} \\
&\mathcal{E} ::= \mathsf{E}\mathcal{T} && \text{(pure Existence)} \\
&\mathcal{R} ::= \mathcal{A} \mid \mathcal{B} \mid \mathcal{D} \mid \mathcal{E} && \text{(Reducible term)} \\
&\mathcal{T} ::= \mathcal{V} \mid \mathcal{N} \mid \mathcal{R} && \text{(Term)}
\end{aligned}
$$

As an example, $\lambda \mathsf{v}_1.\mathsf{if}(\mathsf{v}_1,\mathsf{v}_2,\mathsf{T})$ is a term:

$$\lambda \mathsf{v}_1.\mathsf{if}(\mathsf{v}_1,\mathsf{v}_2,\mathsf{T}) \in \mathcal{T}$$

Let $\mathcal{C}$ denote the set of closed terms (*i.e.*, terms without free variables). As examples, we have $\lambda \mathsf{v}_1.\mathsf{v}_1 \in \mathcal{C}$
$\lambda \mathsf{v}_1.\mathsf{v}_2 \notin \mathcal{C}$

2.2 The syntax expressed in ZFC

We now translate the BNF-definition of the previous section into ZFC. Define

$$
\begin{aligned}
\mathsf{v}_i &\doteq \langle 0, i \rangle \\
\mathsf{T} &\doteq \langle 1 \rangle \\
\lambda x.y &\doteq \langle 2, x, y \rangle \\
xy &\doteq \langle 3, x, y \rangle \\
\mathsf{if}(x,y,z) &\doteq \langle 4, x, y, z \rangle \\
x\|y &\doteq \langle 5, x, y \rangle \\
\mathsf{E}x &\doteq \langle 6, x \rangle
\end{aligned}
$$

Let ω be the set of finite ordinals (*i.e.*, the set of natural numbers). Let $\mathcal{V}, \mathcal{U}, \mathcal{F}, \mathcal{N}, \mathcal{A}, \mathcal{B}, \mathcal{D}, \mathcal{E}, \mathcal{R}$, and $\mathcal{T}$ be the smallest sets such that

---

[1] Readers with a λ-calculus background should note that M is an applied ("impure") λ-calculus because of the ur-element and that the definition of 'normal form' is highly non-standard.

$$
\begin{aligned}
&i \in \omega && \Rightarrow \mathsf{v}_i && \in \mathcal{V} \text{ (object Variable)} \\
&x = \mathsf{T} && \Rightarrow x && \in \mathcal{U} \text{ (Urelement)} \\
&x \in \mathcal{V} \wedge y \in \mathcal{T} && \Rightarrow \lambda x.y && \in \mathcal{F} \text{ (Function)} \\
&x \in \mathcal{U} \cup \mathcal{F} && \Rightarrow x && \in \mathcal{N} \text{ (Normal form)} \\
&x, y \in \mathcal{T} && \Rightarrow xy && \in \mathcal{A} \text{ (functional Application)} \\
&x, y, z \in \mathcal{T} && \Rightarrow \mathsf{if}(x,y,z) && \in \mathcal{B} \text{ (Branch)} \\
&x, y \in \mathcal{T} && \Rightarrow x \| y && \in \mathcal{D} \text{ (Parallel disjunction)} \\
&x \in \mathcal{T} && \Rightarrow \mathsf{E}x && \in \mathcal{E} \text{ (pure Existence)} \\
&x \in \mathcal{A} \cup \mathcal{B} \cup \mathcal{D} \cup \mathcal{E} && \Rightarrow x && \in \mathcal{R} \text{ (Reducible term)} \\
&x \in \mathcal{N} \cup \mathcal{R} && \Rightarrow x && \in \mathcal{T} \text{ (Term)}
\end{aligned}
$$

2.3 Semantics

The semantics of M is a mathematical description of what a computer is supposed to do with terms t.

The semantics of M only considers closed terms $t \in \mathcal{C}$. Given a closed term t, a computer is supposed to *reduce* t, *i.e.*, to transform t according to certain *reduction rules* until t is transformed into normal form, if possible. Hence, the semantics of M is a function f_* of type

$$f_* : \mathcal{C} \to \mathcal{N}$$

where $f_*(t)$ is defined iff t can be reduced to normal form. The function f_* is *partial* in the sense that it is not defined for all $t \in \mathcal{C}$.

2.4 Reduction steps

A computer is supposed to reduce t one *step* at a time, *i.e.*, to compute a *reduction sequence*

$$t_0 \to t_1 \to t_2 \to \cdots$$

of closed terms such that $t_0 = t$ and $t_{i+1} = f_1(t_i)$ where $f_1 : \mathcal{C} \to \mathcal{C}$ defines what to do in one step. If $t_i \in \mathcal{N}$ for no i then $f_*(t)$ is undefined. If $t_i \in \mathcal{N}$ for some i then $f_*(t) = t_i$. The definition of $f_1(t)$ reads:

$$f_1(t) = \begin{cases} t & \text{if } t \in \mathcal{N} \\ f_{\mathcal{A}}(t) & \text{if } t \in \mathcal{A} \\ f_{\mathcal{B}}(t) & \text{if } t \in \mathcal{B} \\ f_{\mathcal{D}}(t) & \text{if } t \in \mathcal{D} \\ f_{\mathcal{E}}(t) & \text{if } t \in \mathcal{E} \end{cases}$$

The function f_1 is *total* in the sense that $f_1(t)$ is defined for all $t \in \mathcal{C}$.

As an example, the reduction sequence of T is

$$\mathsf{T} \to \mathsf{T} \to \mathsf{T} \to \cdots$$

so $f_*(\mathsf{T}) = \mathsf{T}$.

2.5 Branch

The function $f_\mathcal{B}$ is defined as follows for all $x, y, z \in \mathcal{C}$:

$$f_\mathcal{B}(\mathsf{if}(x,y,z)) = \begin{cases} y & \text{if } x \in \mathcal{U} \\ z & \text{if } x \in \mathcal{F} \\ \mathsf{if}(f_1(x),y,z) & \text{if } x \in \mathcal{R} \end{cases}$$

As an example, the reduction sequence of $\mathsf{if}(\mathsf{if}(\mathsf{T},\mathsf{T},\mathsf{T}), \lambda v_1.v_1, \mathsf{T})$ is

$$\mathsf{if}(\mathsf{if}(\mathsf{T},\mathsf{T},\mathsf{T}), \lambda v_1.v_1, \mathsf{T}) \to \mathsf{if}(\mathsf{T}, \lambda v_1.v_1, \mathsf{T}) \to \lambda v_1.v_1 \to \lambda v_1.v_1 \to \cdots$$

so $f_*(\mathsf{if}(\mathsf{if}(\mathsf{T},\mathsf{T},\mathsf{T}), \lambda v_1.v_1, \mathsf{T})) = \lambda v_1.v_1$.

2.6 Application

For all $a, b \in \mathcal{T}$ and $v \in \mathcal{V}$ define $\langle a\,|v{:=}b\,\rangle$ as the result of replacing all free occurrences of v in a by b with suitable renaming of bound variables to avoid variable clashes. We shall assume that renaming is done in some deterministic way which we shall not specify any further. Now define apply: $\mathcal{F} \times \mathcal{T} \to \mathcal{T}$ such that

$$\mathrm{apply}(\lambda v.a, b) = \langle a\,|v{:=}b\,\rangle$$

The function $f_\mathcal{A}$ is defined as follows for all $x, y \in \mathcal{C}$:

$$f_\mathcal{A}(xy) = \begin{cases} x & \text{if } x \in \mathcal{U} \\ \mathrm{apply}(x,y) & \text{if } x \in \mathcal{F} \\ f_1(x)y & \text{if } x \in \mathcal{R} \end{cases}$$

As an example, the reduction sequence of $(\lambda v_1.v_1 v_1)(\lambda v_1.v_1 v_1)$ is

$$(\lambda v_1.v_1 v_1)(\lambda v_1.v_1 v_1) \to (\lambda v_1.v_1 v_1)(\lambda v_1.v_1 v_1) \to \cdots$$

so $f_*((\lambda v_1.v_1 v_1)(\lambda v_1.v_1 v_1))$ is undefined.

2.7 Parallel disjunction

The function $f_\mathcal{D}$ is defined as follows for all $x, y \in \mathcal{C}$:

$$f_\mathcal{D}(x\|y) = \begin{cases} \mathsf{T} & \text{if } x \in \mathcal{U} \text{ or } y \in \mathcal{U} \\ \lambda v_1.(xv_1)\|(yv_1) & \text{if } x \in \mathcal{F} \text{ and } y \in \mathcal{F} \\ f_1(x)\|f_1(y) & \text{otherwise} \end{cases}$$

Hence, $x\|y$ is reduced by reducing x and y in parallel until one of them reduces to T or both of them reduce to functions.

2.8 Existence

The function $f_\mathcal{E}$ is defined as follows for all $x \in \mathcal{C}$:

$$\begin{aligned} S &\doteq \lambda \mathsf{v}_1.\lambda \mathsf{v}_2.\lambda \mathsf{v}_3.\mathsf{v}_1\mathsf{v}_3(\mathsf{v}_2\mathsf{v}_3) \\ K &\doteq \lambda \mathsf{v}_1.\lambda \mathsf{v}_2.\mathsf{v}_1 \\ B &\doteq \lambda \mathsf{v}_1.\lambda \mathsf{v}_2.\lambda \mathsf{v}_3.\mathsf{if}(\mathsf{v}_1,\mathsf{v}_2,\mathsf{v}_3) \\ D &\doteq \lambda \mathsf{v}_1.\lambda \mathsf{v}_2.\mathsf{v}_1 \| \mathsf{v}_2 \\ E &\doteq \lambda \mathsf{v}_1.\mathsf{E}\mathsf{v}_1 \\ A &\doteq \lambda \mathsf{v}_1.\mathsf{E}\lambda \mathsf{v}_2.\mathsf{E}\lambda \mathsf{v}_3.\mathsf{v}_1(\mathsf{v}_2\mathsf{v}_3) \\ f_\mathcal{E}(\mathsf{E}x) &= xS \| xK \| x\mathsf{T} \| xB \| xD \| xE \| Ax \end{aligned}$$

Essentially, $\mathsf{E}x$ is reduced by reducing xy for all $y \in \mathcal{C}$ in parallel. $\mathsf{E}x$ reduces to T iff xy reduces to T for some $y \in \mathcal{C}$. $\mathsf{E}x$ is a very weak quantifier. At a later stage, map theory is constructed by adding a different and much stronger quantifier.

3 A theory of M

3.1 The M-computer

One may think of an implementation of the M programming language as an *M-computer* with two lamps and a keyboard. The two lamps are labeled T and $\mathcal{F}$, and the keyboard has the symbols used in the BNF-definition in Section 2.1. When a user enters a closed term on the keyboard, the computer starts reducing the term. If the term reduces to a normal form, then the T and $\mathcal{F}$ lamp lights if the result is in $\mathcal{U}$ and $\mathcal{F}$, respectively[2].

We shall say that a closed term t is a *true*, *function*, or *bottom* term if $f(t) \in \mathcal{U}$, $f(t) \in \mathcal{F}$, or $f(t)$ is undefined, respectively. Hence, if the M-computer receives a true term, it lights the T lamp, if it receives a function term, it lights the $\mathcal{F}$ lamp, and if it receives a bottom term then it will work indefinitely without lighting either lamp.

Hence, an ordinary user of the M-computer can use it to verify that a term is a true or a function term. It takes a clairvoyant user to verify that a term is a bottom term.

We shall say that two closed terms r and s are *root equivalent*, written $r \approx s$, if they are both true, both function, or both bottom terms. Root equivalence is not computable because it is undecidable whether or not a term is a bottom term.

---

[2] Readers with a background in λ-calculus will note that this is very different from a typical implementation of pure lambda calculus: in pure λ-calculus, the result of a computation is a normal form; in M, the result of a computation is binary and the user cannot access the normal form

3.2 Equivalence

We shall say that two closed terms r and s are *equivalent*, written $r \equiv s$, if $tr \approx ts$ for all closed terms t. Hence, two terms are equivalent if they behave the same on the M-computer in any context t.

From the point of view of MT, $r \equiv s$ expresses equality.

In the present paper, we use $r = s$ to denote equality in ZFC. Hence, for closed terms r and s, $r = s$ expresses that r and s are the same term and $r \equiv s$ expresses that r and s are equal in MT. In the present paper, $r \equiv s$ is an equivalence relation in ZFC.

Some papers on MT are based on MT rather than ZFC. In those papers, equality in MT is still written $x \equiv y$ whereas $x = y$ is used for quite another purpose: to denote equality modulo identifications. So in those papers, $x \equiv y$ is equality whereas $x = y$ merely is an equivalence relation.

We shall refer to the equivalence classes of the class division $\mathcal{C}/\equiv$ as *maps*[3].

3.3 Elementary properties

Let $x \propto \langle y|v{:=}z\rangle$ denote that x is identical to $\langle y|v{:=}z\rangle$ except for renaming of bound variables and define

$$\begin{aligned}
\bot &\doteq (\lambda v_1.v_1v_1)(\lambda v_1.v_1v_1) \\
x \circ y &\doteq (\lambda v_1.\lambda v_2.\lambda v_3.v_1(v_2v_3))xy \\
? &\doteq \lambda v_1.\mathsf{if}(v_1, \mathsf{T}, \bot) \\
x \to y &\Leftrightarrow \mathsf{if}(x, y, \mathsf{T}) \equiv \mathsf{if}(x, \mathsf{T}, \mathsf{T})
\end{aligned}$$

We have that $\bot$ is a bottom term, $x \circ y$ is functional composition, ? is a function which maps T to T and anything else to $\bot$, and $x \to y$ is one way to express that $x \equiv \mathsf{T}$ implies $y \equiv \mathsf{T}$ in MT. We state without proof that the following hold for all terms $x, y, z \in \mathcal{T}$ and all variables $v \in \mathcal{V}$:

---

[3] So the set of maps is countable until further. The notion of a map is going to be generalized when we introduce MT.

$$x \equiv y \Rightarrow x \equiv z \Rightarrow y \equiv z^4$$
$$x \equiv y \Rightarrow \lambda v.x \equiv \lambda v.y$$
$$x \equiv y \Rightarrow zx \equiv zy$$

$$\mathsf{if}(\mathsf{T}, y, z) \equiv y$$
$$\mathsf{if}(\lambda v.x, y, z) \equiv z$$
$$\mathsf{if}(\bot, y, z) \equiv z$$

$$\mathsf{T}y \equiv \mathsf{T}$$
$$(\lambda v.y)z \equiv x \quad \text{if } x \propto \langle y|v{:=}z\rangle$$
$$\bot y \equiv \bot$$

$$\mathsf{T}\|x \equiv \mathsf{T}$$
$$x\|\mathsf{T} \equiv \mathsf{T}$$
$$(\lambda v.x)\|(\lambda v.y) \equiv \lambda v.(x\|y)$$
$$(\lambda v.x)\|\bot \equiv \bot$$
$$\bot\|(\lambda v.x) \equiv \bot$$
$$\bot\|\bot \equiv \bot$$

$$\mathsf{E}\mathsf{T} \equiv \mathsf{T}$$
$$\mathsf{E}\bot \equiv \bot$$
$$\mathsf{E}(x \circ y) \rightarrow \mathsf{E}x$$
$$\mathsf{E}(? \circ x) \equiv \mathsf{E}x$$

The statements above are the elementary axioms and inference rules of MT.

3.4 Quartum non datur

The rule of Quartum Non Datur (QND) reads:

$$x\mathsf{T} \equiv y\mathsf{T} \Rightarrow x(\lambda v_1.v_2v_1) \equiv y(\lambda v_1.v_2v_1) \Rightarrow x\bot \equiv y\bot \Rightarrow xv_2 \equiv yv_2$$

The rule says that any closed term is a true, function, or bottom term, there is no fourth possibility. QND allows to prove a lemma by cases. Now define:

$$\mathsf{F} \doteq \lambda v_1.\mathsf{T}$$
$$\neg x \doteq \mathsf{if}(x, \mathsf{F}, \mathsf{T})$$
$$x \wedge y \doteq \mathsf{if}(x, \mathsf{if}(y, \mathsf{T}, \mathsf{F}), \mathsf{if}(y, \mathsf{F}, \mathsf{F}))$$

We use F to represent falsehood and $\neg x$ and $x \wedge y$ express negation and conjunction in M (and, thereby, in MT). We use $\wedge$ for conjunction in both MT and ZFC, but which one is meant should be clear from the context. As an example, in the fact

---

[4] We take $\Rightarrow$ to be right associative so that $A \Rightarrow B \Rightarrow C$ means $A \Rightarrow (B \Rightarrow C)$ which is equivalent to $A \wedge B \Rightarrow C$.

$$(x \wedge y) \equiv \mathsf{T} \Leftrightarrow (x \equiv \mathsf{T}) \wedge (y \equiv \mathsf{T})$$

the leftmost $\wedge$ is part of an MT term whereas the rightmost $\wedge$ is part of a ZFC formula. Note, by the way, that conjunction lives at the level of terms in MT and at the level of formulas in ZFC. As we shall see later, the same holds true for quantifiers. The elementary properties listed in Section 3.3 allow to prove *e.g.*

$$\begin{array}{|l|l|l|l|} \hline \neg \mathsf{T} \equiv \mathsf{F} & \mathsf{T} \wedge \mathsf{T} \equiv \mathsf{T} & \mathsf{T} \wedge \mathsf{F} \equiv \mathsf{F} & \mathsf{T} \wedge \bot \equiv \bot \\ \neg \mathsf{F} \equiv \mathsf{T} & \mathsf{F} \wedge \mathsf{T} \equiv \mathsf{F} & \mathsf{F} \wedge \mathsf{F} \equiv \mathsf{F} & \mathsf{F} \wedge \bot \equiv \bot \\ \neg \bot \equiv \bot & \bot \wedge \mathsf{T} \equiv \bot & \bot \wedge \mathsf{F} \equiv \bot & \bot \wedge \bot \equiv \bot \\ \hline \end{array}$$

QND is required to prove more general statements like $v_1 \wedge v_2 \equiv v_2 \wedge v_1$ and $\neg \neg \neg v_1 \equiv \neg v_1$. QND is unable to prove $\neg \neg v_1 \equiv v_1$ for the simple reason that $\neg \neg v_1 \equiv v_1$ does not hold in general (as an example, $\neg \neg \lambda v_1.\bot \equiv \neg \mathsf{T} \equiv \mathsf{F} \equiv \lambda v_1.\mathsf{T} \not\equiv \lambda v_1.\bot$).

3.5 The Scott order

Now define

$$\begin{aligned} \mathsf{Y} &\doteq \lambda v_1.(\lambda v_2.v_1(v_2 v_2))(\lambda v_2.v_1(v_2 v_2)) \\ x \downarrow y &\doteq \mathsf{if}(x, \mathsf{if}(y, \mathsf{T}, \bot), \mathsf{if}(y, \bot, \lambda z.xz \downarrow yz)) \\ x \preceq y &\Leftrightarrow x \equiv x \downarrow y \end{aligned}$$

The elementary properties allow to prove $\mathsf{Y}x \equiv x(\mathsf{Y}x)$ showing that Y is a fixed point operator. Having a fixed point operator makes recursive definitions like the second definition above permissible as any recursive definition can be translated to a non-recursive one using Y.

In the 'standard model' of MT, $x \downarrow y$ is the infimum (greatest lower bound) of x and y w.r.t. the Scott order whereas $x \preceq y$ expresses the Scott order itself.

Having defined the Scott order $x \preceq y$ we can state two more rules about maps, one which says that all maps are monotonic, and one which says that Y produces minimal fixed points:

$$\begin{aligned} x \preceq y &\Rightarrow zx \preceq zy \\ xy \preceq y &\Rightarrow \mathsf{Y}x \preceq y \end{aligned}$$

In addition, M also satisfies the property of Scott continuity, but that property is not included as an inference rule of MT since that would hinder the introduction of genuine quantifiers (the $\mathsf{E}x$ quantifier is not a genuine quantifier; it can only return T and $\bot$ but not F).

As proved in Berline and Grue (1997), the standard model of full map theory has a generalized property known as κ-Scott continuity where κ is a suitably chosen ordinal. As a special case of κ-Scott continuity, ordinary Scott continuity is identical to ω-Scott continuity where ω is the set of finite ordinals.

3.6 Extensionality

ZFC has the *extensionality* property that the following are equivalent:

1. $x \in z \Leftrightarrow y \in z$ for all sets z.
2. $z \in x \Leftrightarrow z \in y$ for all sets z.

M happens to have a similar property which says that the following slightly more complicated statements are equivalent:

1. $zx \approx zy$ for all $z \in \mathcal{C}$.
2. $xz_1 \cdots z_n \approx yz_1 \cdots z_n$ for all $n \in \mathbf{N}$ and all $z_1, \ldots, z_n \in \mathcal{C}$.

Statement (1) above is the definition of $x \equiv y$ and (2) follows from (1) according to the elementary properties of M. Even though it may not be evident, the following *rule of extensionality* expresses that (1) follows from (2):

$$\neg\neg(xu) \equiv \neg\neg(yu) \Rightarrow xuv \equiv x(zuv) \Rightarrow yuv \equiv y(zuv) \Rightarrow xu \equiv yu$$

Above, $u, v \in \mathcal{V}$ must be distinct variables and $x, y, z \in \mathcal{T}$ must be terms in which u and v do not occur free.

Extensionality allows to prove $x \downarrow x \equiv x$, $x \downarrow y \equiv y \downarrow x$, and $x \downarrow (y \downarrow z) \equiv (x \downarrow y) \downarrow z$ (cf. Grue (2002b)). These results in turn allow to prove $x \preceq x$, $x \preceq y \Rightarrow y \preceq x \Rightarrow x \equiv y$, and $x \preceq y \Rightarrow y \preceq z \Rightarrow x \preceq z$.

4 Models

4.1 Trees

As mentioned in Section 3.6, two maps x and y are equal iff

$$xz_1 \cdots z_n \approx yz_1 \cdots z_n \quad \text{for all } n \in \mathbf{N} \text{ and all } z_1, \ldots, z_n \in \mathcal{C}$$

That property may be used as the basis for a useful mental picture of what a map looks like. The mental picture described in the following is a graphical one which we illustrate by 'drawing' the map $I = \lambda x.x$.

Among other, the map I has the following properties:

$$\begin{aligned} \mathsf{I} &\in \mathcal{F} \\ \mathsf{IT} &\in \mathcal{U} \\ \mathsf{IF} &\in \mathcal{F} \\ \mathsf{IF}\bot &\in \mathcal{U} \end{aligned}$$

The information above may be represented by the graphical construction of Fig. 1

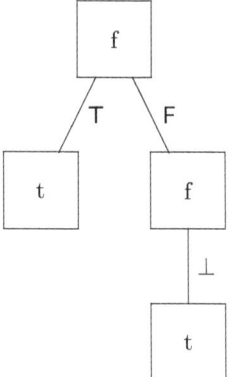

Fig. 1. A drawing of $\lambda x.\mathsf{if}(x\bot, \mathsf{T}, \lambda y.\mathsf{T})$

In the drawing, nodes labeled t and f represent the sets $\mathcal{U}$ and $\mathcal{F}$, respectively. The f in the root represents the $\mathcal{F}$ in $\mathsf{I} \in \mathcal{F}$. The t in the bottom right corner represents the $\mathcal{U}$ in $\mathsf{IF}\bot \in \mathcal{U}$.

In general, in a drawing of a map m, if $mx_1 \cdots x_n \in \mathcal{F}$ then there is a node labeled f and a path from the root to that node labeled $x_1, \ldots, x_n$.

We do not allow downward edges from nodes labeled t, so the rule for nodes labeled t is a bit more complicated: If $mx_1 \cdots x_n \in \mathcal{U}$ and if $mx_1 \cdots x_r \notin \mathcal{U}$ for all $r < n$ then there is a node labeled t and a path from the root to that node labeled $x_1, \ldots, x_n$.

A drawing like the one above only records positive information of form $mx_1 \cdots x_n \in \mathcal{U}$ or $mx_1 \cdots x_n \in \mathcal{F}$. It does not record negative information of form $mx_1 \cdots x_n \equiv \bot$.

The drawing above merely is an approximation of I since it does not record all paths $x_1 \cdots x_n$ for which $\mathsf{I} x_1 \cdots x_n \not\equiv \bot$. The picture above may be seen as a drawing of

$$\lambda x.\mathsf{if}(x\bot, \mathsf{T}, \lambda y.\mathsf{T})$$

since that is the smallest map w.r.t. the Scott order which has the illustrated properties. That map is indeed an *approximation* of I in the sense that

$$\lambda x.\mathsf{if}(x\bot, \mathsf{T}, \lambda y.\mathsf{T}) \preceq \mathsf{I}$$

A full drawing of I would be infinitely large.

4.2 Compact maps

We shall say that a map c is *compact* iff there exists a map χ_c such that $\chi_c d \equiv \mathsf{T}$ iff $c \preceq d$ for all maps d. The compact maps are exactly those which are also compact in the standard model of map theory.

98   Klaus Grue

In drawings, one may restrict edge labels to compact maps without loss of information. From now on, we assume that edge labels are restricted to be compact maps.

In the drawing of the approximation towards I in Fig. 1, the edge labels T, F, and ⊥ are themselves maps and may themselves be drawn instead of referred to. A drawing of T consists of a single node labeled t. A drawing of ⊥ is hard to see since it consists of a tree with no nodes. Fig. 2 shows a drawing of F.

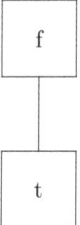

Fig. 2. A drawing of F

The drawing in Fig. 2 consists of an f-node and a t-node connected by an edge labeled ⊥. The ⊥ can (or, rather, cannot) be seen next to the label.

All compact maps of M happen to have finitely large drawings[5]. This holds true even if one insists that all edge labels have to be drawn themselves. Compact maps of full MT need not have finitely large drawings.

4.3 Prime maps

The map drawn of Fig. 1 may be seen as the least upper bound of the two maps in Fig. 3.

We shall say that a map is *prime* if it has finitely many nodes and each node has at most one downward edge. The prime maps are exactly those which are also prime in the standard model of map theory. As a technicality we require prime maps to have at least one node so that we expel ⊥ from the society of prime maps. The two maps above constitute a prime factoring of the map drawn in Fig. 1.

4.4 Coherent maps

We shall say that two maps x and y are *coherent*, written $x \mathrel{\circ} y$, if they have an upper bound, *i.e.*, if there exists a map z such that $x \preceq z$ and $y \preceq z$. The two prime maps in Section 4.3 are coherent.

---

[5] More precisely: for all compact maps c there exists a finitely large drawing which represents c. Drawings may contain redundant information and a compact map can easily have many finite as well as many infinite drawings.

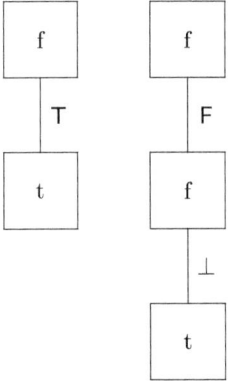

Fig. 3. A prime factoring of the map from Fig. 1

If two maps x and y have a common upper bound z then they also have a least upper bound $\mathrm{lub}(x, y, z)$ where

$$\mathrm{lub}(x, y, z) \doteq \mathrm{if}(z, x : y : \mathsf{T}, \neg x : \neg y : \lambda u.\mathrm{lub}(xu, yu, zu))$$
$$x : y \doteq \mathrm{if}(x, y, \bot)$$

For coherent maps x and y, let $x \uparrow y$ denote the least upper bound of x and y. For any finite set S of pairwise coherent maps let $\uparrow S$ denote the least upper bound of all the maps in S.[6]

4.5 Model building kit

For compact c and maps m let $c \mapsto m$ denote $\lambda x.\mathrm{if}(\chi_c x, m, \bot)$. Furthermore, let t and f denote T and $\lambda x.\bot$, respectively. Fig. 4 illustrates the capabilities of t, f, $\uparrow$, and $\mapsto$.

We have, $t = \mathsf{T}$, $\uparrow \emptyset = \bot$, and $(\uparrow \emptyset) \mapsto t = \mathsf{F}$, so we may also express the rightmost tree of Fig. 4 as

$$t \mapsto t \uparrow ((\uparrow \emptyset) \mapsto t) \mapsto (\uparrow \emptyset) \mapsto t$$

Now define $x \hookrightarrow y = (\uparrow x) \mapsto y$. This allows to express the rightmost tree of Fig. 4 as the least upper bound of

$$\{\{t\} \hookrightarrow t, \{\emptyset \hookrightarrow t\} \hookrightarrow \emptyset \hookrightarrow t\}$$

4.6 Model building

All prime maps can be expressed using only t, f, $x \hookrightarrow y$, and the ability to form finite sets of pairwise coherent maps. Furthermore, all maps can be expressed as supprema of prime maps. Now define:

---

[6] That least upper bound is guaranteed to exist in the 'standard model' of MT

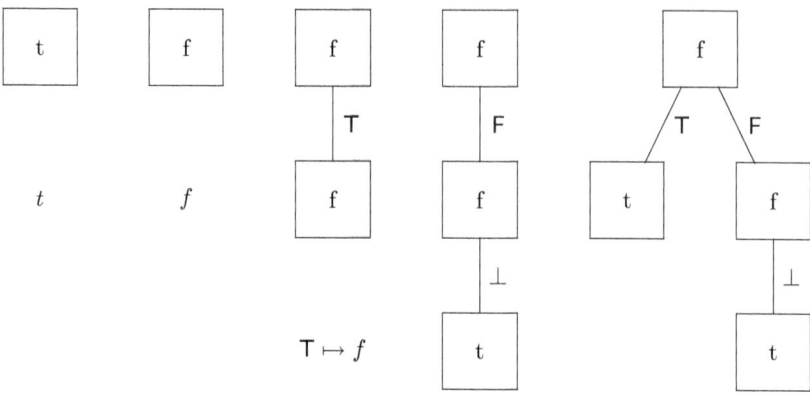

Fig. 4.

$$t = \langle 0 \rangle$$
$$f = \langle 1 \rangle$$
$$x \hookrightarrow y = \langle 2, x, y \rangle$$

Then one may define a set P which models prime maps and a set C which models compact maps by defining P and C as the smallest sets such that

1. P contains t and f and also contains $x \hookrightarrow y$ for all $x \in C$ and $y \in P$.
2. C is the set of finite sets of coherent elements of P.

In (2) above, one needs to know which elements of P are coherent. The cases where elements of P are coherent are the following:

1. $t \bigcirc t$.
2. $f \bigcirc f$.
3. $f \bigcirc x \hookrightarrow y$ for all $x \in C$ and $y \in P$.
4. $x \hookrightarrow y \bigcirc f$ for all $x \in C$ and $y \in P$.
5. $x \hookrightarrow y \bigcirc x' \hookrightarrow y'$ iff $x \not\bigcirc x' \vee y \bigcirc y'$ for $x, x' \in C$ and $y, y' \in P$.

In (5) above, one needs to know which elements of C are coherent. Two elements x and y of C are coherent when all elements of x are coherent with all elements of y.

The rules above allow to define P, C, and coherence by mutual recursion which allows to construct a model of the M programming language and the axioms and inference rules stated so far.

To get a model of full map theory, one has to use a large cardinal κ and define C as the set of sets with cardinality less than κ of coherent elements of P (to be precise, one has to use an inaccessible ordinal σ and then choose κ as a regular ordinal larger than σ). A consistency proof of map theory based on these ideas can be found in Berline and Grue (1997).

5 Quantification

5.1 Choices to be made

Map theory (MT) is an extension of the theory from Section 3. We obtain MT by adding a quantifier to the syntax and by extending the collection of axioms. When doing so, two choices must be made.

First, one has to decide which kind of quantifier to add. The obvious possibilities are the universal quantifier $\forall$, the existential quantifier $\exists$, or Hilberts epsilon operator ε.

Second, one has to decide which domain to quantify over. The quantifier of ZFC quantifies over all sets, so the obvious choice for MT would be to quantify over all maps. That turns out to be a bad idea, however.

5.2 Choice of quantifier

Concerning the choice of quantifier, it does not matter whether one chooses $\forall$ or $\exists$ since each one is easy to define from the other. So let us rule out $\exists$.

One can define $\forall$ from ε, but the opposite is not possible, so the choice between $\forall$ and ε does matter.

Since ε is stronger than $\forall$, ε is the most obvious candidate. Including ε in MT gives the axiom of choice for free. Furthermore, the combination of ε and the fixed point operator Y allows to give a particularly easy proof of the well-ordering theorem: one can simply well-order any set by recursive application of ε in a very natural way. One can say more pro and cons ε, however.

5.3 New function letters

The ε quantifier has a less celebrated property, which may be of increasing importance in the future: It makes it much simpler to introduce new functions in a theory. As an example, in Mendelsons system S of Peano arithmetic (Mendelson, 1987), one can prove the existence and uniqueness of quotient and remainder:

$$\forall a, b \exists_1 q, r \colon b \neq 0 \wedge a = bq + r \wedge r < b \vee b = 0 \wedge q = 0 \wedge r = a$$

According to the meta-theorem of new constants and function letters (c. f.. Mendelson (1987)), the theorem above allows to construct an extended system S' with two new function letters, x/y and $x\%y$, and a new axiom,

$$\forall a,b\colon b \neq 0 \wedge a = b(a/b) + a\%b \wedge a\%b < b \vee b = 0 \wedge a/b = 0 \wedge a\%b = a$$

The meta-theorem of new function letters says that if S is consistent then S' is also consistent and is a conservative extension of S, so that one may just as well work in S'. That works fine for human mathematicians, but it is cumbersome to change theory when doing machine verification. Having ε one may simply define x/y and $x\%y$ thus:

$$a/b = \varepsilon q \exists r\colon a = bq + r \wedge r < b$$
$$a\%b = \varepsilon r \exists q\colon a = bq + r \wedge r < b$$

These definitions are easy to deal with in mechanical proof systems.

5.4 Problems with ε

If one includes ε in MT then the Axiom of Choice (AC) becomes provable in MT. So objections against AC may be taken as objections against ε.

AC was first formulated by Zermelo even though it had been in use before. Zermelo used AC for proving Zorn's lemma, and it is said that Zermelo was taunted for AC since it seemed too easy to prove lemmas if one was allowed to invent new axioms for each new lemma.

The objections against AC seems to be:

- AC is dangerous. AC was formulated during the third fundamental crisis where new dubious axioms were probably not particularly welcome. As ZF and ZFC have been shown to be equiconsistent, however, this objection is no longer valid.
- AC is superflous. At the time of Zermelo, it must have been the hope to be able to prove Zorns lemma without the use of new axioms. As the independence of AC from ZF has been shown, this objection is no longer valid either.
- AC has peculiar consequences like the ability to divide a sphere into two spheres of the same size of the original. This seems to be inevitable for axioms about infinities. Even the axiom of infinity has a peculiar consequence: together with the power set axiom it entails that there exist infinities of different sizes.
- AC is unnecessary. One can develop substantial amounts of mathematics without AC. The same objection can be raised against e.g. the law of excluded middle or the axiom of replacement, and it is of course a matter of taste which assumptions to include in a theory.

It is the purpose of MT to be a convenient foundation for the working mathematician, and for that reason ε has been included in it. The benefits are that one can make definitions like those in Section 5.3 easily and one can use AC freely. The drawback is that AC is hardwired into MT so that one cannot easily drop it as one can with AC in ZFC.

5.5 Strict versus lazy quantifiers

In MT, $\forall$ lives at the level of terms and not at the level of well-formed formulas. In MT, $\forall$ is a function which, when applied to an argument p as in $\forall p$, tests whether or not $px \equiv \mathsf{T}$ for all x in the domain of quantification.

Let Φ denote the domain of quantification. The domain of quantification is going to be defined in Section 6. Section 6.1 explains why Φ should not be taken to be the set of all maps. Section 6 chooses a Φ which is neither the set of all maps, nor the empty set.

Like any other function of MT, $\forall$ must be monotonic such that $p \preceq q$ implies $\forall p \preceq \forall q$. The two most obvious ways of achieving monotonicity are the following *strict* and *lazy* quantifiers:

$$\forall_{\text{strict}} p = \begin{cases} \mathsf{T} \text{ if } px \equiv \mathsf{T} \text{ for all } x \in \Phi \\ \bot \text{ if } px \equiv \bot \text{ for some } x \in \Phi \\ \mathsf{F} \text{ otherwise} \end{cases}$$

$$\forall_{\text{lazy}} p = \begin{cases} \mathsf{T} \text{ if } px \equiv \mathsf{T} \text{ for all } x \in \Phi \\ \mathsf{F} \text{ if } px \not\equiv \mathsf{T}, \bot \text{ for some } x \in \Phi \\ \bot \text{ otherwise} \end{cases}$$

The choice operator ε is also a function. When applied to an argument as in εp, it returns an $x \in \Phi$ such that $px \equiv \mathsf{T}$. There has to be two exceptions, however. Firstly, if $px \not\equiv \mathsf{T}$ for all $x \in \Phi$ then ε has to be excused for not returning an $x \in \Phi$ such that $px \equiv \mathsf{T}$. Secondly, like any other function, ε has to be monotonic which puts further restrictions on ε. A strict ε has the following properties:

- $\varepsilon p \equiv \bot$ if $px \equiv \bot$ for some $x \in \Phi$.
- $\varepsilon p \in \Phi$ if $px \not\equiv \bot$ for all $x \in \Phi$.
- $p(\varepsilon p) \equiv \mathsf{T}$ if $px \not\equiv \bot$ for all $x \in \Phi$ and $px \equiv \mathsf{T}$ for some $x \in \Phi$.

A lazy ε which satisfies

$$p(\varepsilon p) \equiv \mathsf{T} \text{ if } px \equiv \mathsf{T} \text{ for some } x \in \Phi \tag{1}$$

is not tenable. To see that let $a, b \in \Phi$ be compact maps which satisfy $a \not\preceq b$. If (1) holds then

- $a \preceq \varepsilon \chi_a \preceq \varepsilon \lambda x.\mathsf{T}$
- $b \preceq \varepsilon \chi_b \preceq \varepsilon \lambda x.\mathsf{T}$

contradicting $a \not\preceq b$.

It is possible to construct choice operators which are more lazy than the strict one, but they are complicated and difficult to work with. For that reason, we include the strict ε in MT. That settles the choice of universal quantifier in favor of the strict one so that $\forall$ from now on denotes $\forall_{\text{strict}}$.

5.6 The syntax of Map Theory

The syntax of map theory is the same as the one defined in Section 2.1 except that ε is added as a new construct:

$$\mathcal{H} ::= \varepsilon \qquad \text{(Hilbert epsilon)}$$
$$\mathcal{T} ::= \mathcal{V} \mid \mathcal{N} \mid \mathcal{R} \mid \mathcal{H} \text{ (Term)}$$

Section 6 defines the domain Φ of quantification and also defines a map ϕ with the following property:

$$\phi x = \begin{cases} \mathsf{T} & \text{if } x \in \Phi \\ \bot & \text{otherwise} \end{cases}$$

We shall refer to elements of Φ as *wellfounded maps*. Section 6 defines ϕ using the syntax of Map Theory given above. Now define

$$\begin{aligned}
!x &\doteq \mathsf{if}(x, \mathsf{T}, \mathsf{T}) \\
\exists &\doteq \lambda p.\ \neg\neg(p(\varepsilon p)) \\
\exists x\colon P &\doteq \exists(\lambda x.\ P) \\
\forall x\colon P &\doteq \neg \exists x\colon \neg P \\
\forall &\doteq \lambda p.\ \forall x\colon px \\
\varepsilon x\colon P &\doteq \varepsilon(\lambda x.\ P)
\end{aligned}$$

We have that $!x \equiv \mathsf{T}$ iff x is *Boolean*, i.e., iff $x \not\equiv \bot$ and that $\forall p \equiv \mathsf{T}$ if $px \equiv \mathsf{T}$ for all wellfounded x. The two negation signs in the definition of $\exists$ ensures that $\exists p$ equals T or F or $\bot$. The constructs $\exists x\colon P$, $\forall x\colon P$, and $\varepsilon x\colon P$ are introduced for notational convenience.

5.7 Axioms of quantification

The axioms of quantification of map theory read:

$$\begin{aligned}
\phi(\varepsilon p) &\equiv \forall x\colon !(px) \\
!(\forall p) &\equiv \forall x\colon !(px) \\
\forall p &\to \phi x \to px \\
\varepsilon p &\equiv \varepsilon x\colon \phi x \wedge px
\end{aligned}$$

The first axiom says that εp is wellfounded iff px is Boolean for all wellfounded x. The second says that $\forall p$ is Boolean under the same condition. The third axiom says that if py is true for all wellfounded y and x is wellfounded then px is true. The last axiom expresses *Ackermann's axiom* (Felgner (1976), p.244).

The definition of Map Theory is now complete except that the definition of ϕ is not yet stated.

6 Choice of domain

6.1 Size of domain

The choice of domain Φ remains. To simplify the discussion, we shall consider the properties of $\forall$ instead of ε for each possible choice of Φ.

If we take Φ to be the set of all maps then, among other, $\bot \in \Phi$. In this case, monotonicity gives us

$$px \equiv \mathsf{T} \text{ for all } x \in \Phi \text{ if and only if } p\bot \equiv \mathsf{T}.$$

so the quantifier becomes trivial:

$$\forall \equiv \lambda p.\neg\neg(p\bot)$$

Hence, it is reasonable to require $\bot \notin \Phi$. The other extreme would be to let Φ be the empty set, but then we would have

$$\forall \equiv \lambda p.\mathsf{T}$$

So Φ should neither be too big, nor too small. To ensure that Φ is of any use, Φ must have an infinite subset Φ' of compact maps such that for all $x, y \in \Phi', x \neq y$ we have

- $x \not\supset y$
- $(x \downarrow y) \notin \Phi$

6.2 Representation of sets

When MT was constructed first time Grue (1992), it was easy to see that Φ should not contain maps like $\bot$ and $\lambda x.\bot$ but could safely contain maps like T and $\lambda x.\lambda y.\mathsf{T}$.

One of the intensions of MT was to build a foundation on top of a programming language which had the same strength as ZFC. For that reason, Φ had to contain a subset Φ' with the properties stated in Section 6.1 with the further property that every set in the universe of ZFC should be representable by an element of Φ'. Furthermore, it should be possible to model the membership relation of ZFC by a function in MT.

To achieve this it is reasonable to consider how to represent sets of ZFC by maps of MT. An obvious choice would be to let a map m represent the set $\{x | mx \equiv \mathsf{T}\}$. Trying that was unfruitful, however.

A representation of sets which turned out to work was the following: let T represent the empty set. Furthermore, if $m \in \Phi$ and $m \neq \mathsf{T}$ then let m represent the set $\{mx | x \in \Phi\}$. This representation happens to work, and it immediately hints at the following two properties for Φ:

- $\mathsf{T} \in \Phi$.
- $x, y \in \Phi \Rightarrow xy \in \Phi$.

So Φ should contain T and be closed under application.

6.3 Wellfoundedness

The axiom of *restriction* or *wellfoundedness* of ZFC hints at another property, namely that elements of Φ should be *wellfounded* in a certain sense.

We shall say that a map g is wellfounded w.r.t. a set G of maps if, for all infinite sequences $x_0, x_1, x_2, \cdots \in G$ there exists a natural number n such that

$$g x_0 x_1 \cdots x_{n-1} \equiv \mathsf{T}$$

For all sets G of maps let G° denote the set of maps that are wellfounded w.r.t. G. The chosen representation of sets together with the axiom of wellfoundedness entail that all elements of Φ should be wellfounded w.r.t. Φ:

$$\Phi \subseteq \Phi^\circ$$

It would be convenient to have $\Phi = \Phi^\circ$. To see that $\Phi = \Phi^\circ$ is not tenable, however, we prove $\mathsf{I} \in \Phi^\circ$ and $\mathsf{I} \notin \Phi$ where $\mathsf{I} = \lambda x.\, x$.

$\mathsf{I} \in \Phi^\circ$: Suppose $x_0, x_1, x_2, \cdots \in \Phi$. We shall find a natural number n such that $\mathsf{I} x_0 x_1 \cdots x_n \equiv \mathsf{T}$. From $x_0 \in \Phi \subseteq \Phi^\circ$ we have $x_0 \in \Phi^\circ$. Hence, there exists a natural number n such that $x_0 x_1 x_2 \cdots x_n \equiv \mathsf{T}$ so $\mathsf{I} x_0 x_1 \cdots x_n \equiv \mathsf{T}$.

$\mathsf{I} \notin \Phi$: Suppose $\mathsf{I} \in \Phi$. We have $\mathsf{I}, \mathsf{I}, \mathsf{I}, \ldots \in \Phi$ so $\mathsf{I} \in \Phi^\circ$ entails that there exists a natural number n such that

$$\underbrace{\mathsf{I}\,\mathsf{I}\cdots\mathsf{I}}_{n} \equiv \mathsf{T}$$

But the left hand side of the equation above equals I regardless of the value of n which gives a contradiction. Hence, $\mathsf{I} \notin \Phi$. This together with $\mathsf{I} \in \Phi^\circ$ gives

$$\Phi \subset \Phi^\circ$$

6.4 Continuity

From $\mathsf{I} \in \Phi^\circ$ and $\mathsf{I} \notin \Phi$ it is obvious that $\Phi = \Phi^\circ$ is not tenable, but there is another reason why one should expect $\Phi = \Phi^\circ$ to have no solutions: one would expect the cardinality of Φ° to be equal to that of the powerset of Φ and, hence, greater than that of Φ. Nevertheless, if M denotes the set of all maps, then the following properties would be convenient:

(6.4.1) $\Phi \subseteq M$
(6.4.2) $\Phi = \Phi^\circ$
(6.4.3) $M = (M \to M) \cup \{\mathsf{T}, \bot\}$

Equation (6.4.3) is not tenable because the set $M \to M$ of functions from M to M has greater cardinality than M. So both (6.4.2) and (6.4.3) are impossible for cardinality reasons.

The usual countermeasure for a domain equation like (6.4.3) is to restrict $M \to M$ to functions which are continuous in some sense. For ordinary models of lambda-calculus and related theories it is customary to use Scott-continuity. For models of map theory one has to go to the generalized notion of κ-Scott-continuity where κ is an ordinal.

So one may try the same on (6.4.2) and restrict $\Phi°$ to continuous functions. But the notion of continuity used in (6.4.2) must be different from that used in (6.4.3) because the identity function I must be continuous in the sense used in (6.4.3) and discontinuous in the sense used in (6.4.2).

6.5 κ-Scott Continuity

The simplest way to explain κ-Scott continuity is as follows: We say that c is a κ-chain if $c \in \kappa \to M$ and $\forall \beta \in \kappa \forall \alpha \in \beta: c(\alpha) \preceq c(\beta)$. We require M to be κ-complete in the sense that every κ-chain c has a supremum $\uparrow c$ in M. Then a function $g \in M \to M$ is κ-continuous if

$$g(\uparrow c) = \uparrow(g \circ c)$$

for all κ-chains c. This is equivalent to ordinary Scott-continuity for $\kappa = \omega$.[7]

If two ordinals κ and κ' have the same co-finality then the notions of κ- and κ'-continuity are identical. Hence, it is reasonable to restrict κ to be regular, i.e., restrict κ to be equal to its own co-finality.

Now let $A \xrightarrow{\kappa} B$ denote the set of κ-Scott continuous functions from A to B. If κ increases, then the notion of κ-Scott continuity becomes weaker and, hence, $A \xrightarrow{\kappa} B$ becomes bigger. Hence, to get a big universe for map theory one should select a large cardinal κ and replace (6.4.3) by the following:

$$(6.4.3') \ M = (M \xrightarrow{\kappa} M) \cup \{\top, \bot\}$$

6.6 Uniform continuity

For all sets S let S^* denote the set of finite lists of elements of S. For $m \in M$ and $x = \langle x_1, \ldots, x_n \rangle \in M^*$ let mx denote $mx_1 \cdots x_n$. For all sets U and V we shall say that U is V-small if the cardinality of U is less than the cardinality of V. Let $\mathcal{P}_V(S)$ denote the set of V-small subsets of S. For all $S \subseteq M$ define $\Uparrow S = \{y \in M | \exists x \in S: x \preceq y\}$. For all $x \in M$ and $S \subseteq M$ define the neighborhood

$$B(x, S) = \{y \in M | \forall s \in S^*: \neg\neg(xs) \preceq \neg\neg(ys)\}.$$

The set $\{B(x,S) | x \in M \wedge S \in \mathcal{P}_\kappa(M)\}$ of neighborhoods generates the κ-Scott topology. A function $g \in M \to M$ is κ-continuous iff

---

[7] κ-Scott-domains are κ-algebraic and continuity may be defined via commutation with suprema of κ-chains as is done e.g. in Krivine (1993) for ω-continuity, in Plotkin (1982) for ω_1-continuity, and in Berline and Grue (1997) for κ-continuity.

$$\forall x{\in}M \forall U{\in}\mathcal{P}_\kappa(M) \exists V{\in}\mathcal{P}_\kappa(M) \forall y{\in}B(x,V) \colon g(y) \in B(g(x),U)$$

We now define that a function $g \in M \to M$ is *uniformly κ-continuous* iff

$$\forall U{\in}\mathcal{P}_\kappa(M) \exists V{\in}\mathcal{P}_\kappa(M) \forall x{\in}M \forall y{\in}B(x,V) \colon g(y) \in B(g(x),U)$$

Let S_κ° denote the set of maps which are wellfounded w.r.t. S and are uniformly κ-continuous. We are now in a position to reformulate (6.4.2): Let σ be an inaccessible ordinal, let κ be a regular cardinal greater than σ. Instead of (6.4.2) we shall require:

$$(6.4.2') \quad \Phi = \Uparrow \Phi_\sigma^\circ$$

6.7 A property of Φ

The treatment given so far hints at a way to construct a model for Map Theory and hints at a way to define Φ. The model is developed in Berline and Grue (1997), but the definition of Φ given there is a bit more convenient to work with: Φ is taken to be the smallest subset of M for which

$$G \in \mathcal{P}_\sigma(\Phi) \Rightarrow (G^\circ \to G) \subseteq \Phi$$

The definition of Φ above is the one we shall formalize in the following.

6.8 Classes

In the following, we use maps m to represent classes $\{x | mx \equiv \mathsf{T}\}$. As mentioned in Section 6.2, sets of ZFC are represented another way, and we use the term *class* in the following to distinguish from ZFC sets. The classes considered in the following are not the same as those of NBG set theory. Now define:

$x \sqsubset y$	$\doteq yx$	(class membership)
$\{x \vert P\}$	$\doteq \lambda x.P$	(class comprehension)
V	$\doteq \{x \vert \mathsf{T}\}$	(universal class)
$\{\mathsf{T}\}$	$\doteq \{x \vert x\}$	(class containing only T)
$\bigcup x{\sqsubset}G \colon H$	$\doteq \{m \vert \mathsf{E}x \colon x \sqsubset G \wedge m \sqsubset H\}$	(Union class)

In $\{x\vert P\}$ and $\bigcup x{\sqsubset}G \colon H$ above, x may occur free in P and H.

6.9 The dual class

In line with Section 6.3, let $G°$ denotes the class of maps which are wellfounded w.r.t. the class G. We have $G \preceq H \Rightarrow H° \preceq G°$ so monotonicity prevents us from defining $G°$ in map theory. But we can define a universal quantifier which quantifies over $G°$ and a set $G° \to H$ of functions with domain $G°$ and range H. To do so we define $g|G$ such that $g|G$ is the function g recursively restricted to the class G.

$$g|G \doteq \text{if}(g, \mathsf{T}, \lambda x. \text{if}(x \sqsubset G, (gx)|G, \bot))$$
$$\forall x \sqsubset G° : P \doteq \forall x : (\lambda x.P)(x|G)$$
$$G° \to H \doteq \{g | \forall x \sqsubset G° : gx \sqsubset H\}$$

Whenever $g \in \Phi$ and $G \subseteq \Phi$ we have $g|G \in G°$. The definition of $\forall x \sqsubset G° : P$ references $\forall$ which already quantifies over wellfounded maps, so $\forall x \sqsubset G° : P$ quantifies over $G°$ whenever $G \subseteq \Phi$. $G° \to H$ is the set of functions with domain $G°$ and range H whenever $G \subseteq \Phi$. We may also define a lazy and a strict union:

$$\biguplus x \sqsubset G° : H \doteq \{m | \text{E}x : \phi x \wedge m \sqsubset (\lambda x.H)(x|G)\}$$
$$\bigcup x \sqsubset G° : H \doteq (\forall x \sqsubset G° : !H) : (\biguplus x \sqsubset G° : H)$$

The $\biguplus$ operator is a lazy union which treats $\bot$ as the empty set. The $\bigcup$ operator is a more strict union which treats $\bot$ as undefined but treats $\lambda x.\bot$ as the empty set.

6.10 The definition of ϕ

To define ϕ we first define C_a and G_a. C_a and G_a are defined such that $C_a, G_a \in \mathcal{P}_\sigma(\Phi)$ for all maps a.

$$C_a \doteq G_a° \to G_a$$
$$G_a \doteq \text{if}(a, \{\mathsf{T}\}, \bigcup x \sqsubset C_{a\mathsf{T}}° : C_{a\mathsf{F}x})$$
$$\phi \doteq \bigcup a \sqsubset V : C_a$$

An inaccessible ordinal σ has the property that if a set S is σ-small then the powerset $\mathcal{P}(S)$ is also σ-small. That ensures that if G_a is σ-small then so is C_a. Furthermore, the union of a σ-small set of σ-small sets is in turn σ-small. That ensures that if $C_{a\mathsf{T}}$ is σ-small and if $C_{a\mathsf{F}x}$ is σ-small for all $x \in C_{a\mathsf{T}}°$ then G_a is σ-small. The strict union operator $\bigcup$ is used in the definition of G_a to avoid problems with maps a for which $a\mathsf{F}b \equiv a$ for some $b \in C_{a\mathsf{T}}°$.

7 Conclusion

The axioms and inference rules of map theory have been introduced. The axiomatic system is considerably simpler than that of Grue (1992). The model of map theory given in Berline and Grue (1997) is supposed to carry over to the formulation given here. The development of ZFC given in Grue (2002b) is also supposed to carry over to the formulation given here. The last two claims, however, are work of the future.

References

C. Berline and K. Grue. A κ-denotational semantics for Map Theory in ZFC+SI. *Theoretical Computer Science*, 179(1–2):137–202, June 1997.

U. Felgner. Choice functions on sets and classes. In *Sets and Classes: On the works by Paul Bernays*, pages 217–255. North-Holland, 1976.

K. Grue. Map theory. *Theoretical Computer Science*, 102(1):1–133, July 1992.

K. Grue. Lambda-calculus as a foundation for mathematics. In C. Anthony Anderson and Michael Zeleny, editors, *Logic, Meaning and Computation : Essays in Memory of Alonzo Church*, volume 305 of *Synthese Library*, pages 289–314, Dordrecht, 2002a. Kluwer Academic Publishers.

K. Grue. Map theory with classical maps. Technical Report 02/21, DIKU, Universitetsparken 1, DK-2100 Copenhagen, Denmark, 2002b.

J.L. Krivine. *Lambda-calculus, types and models*. Ellis & Horwood, 1993.

E. Mendelson. *Introduction to Mathematical Logic*. Wadsworth and Brooks, 3. edition, 1987.

G. Plotkin. A power domain for countable non-determinism. In *Lecture Notes in Computer Science*, volume 140, pages 418–428. ICALP'82, Springer-Verlag, 1982.

ϵ-Style (of) Semantics
An alternative to set-theoretic modelling

Sebastian Bab, Bernd Mahr, Tina Wieczorek

Faculty IV (Electrical Engineering and Computer Sciences)
Berlin Institute of Technology

{bab,mahr,wieczo}@cs.tu-berlin.de

Summary. An alternative approach to modelling with sets is presented. It uses ϵ-structures as semantic domains for set-like concepts. The resulting ϵ-style of semantics allows the modelling of reflexive structures and self-applicable functions and generally admits to express intensional meaning. A term model of the pure λ-calculus is given and the relationship between Aczel's theory of non-wellfounded sets and ϵ-sets is studied. Based on the idea of ϵ-style semantics, $\in_T$-logics have been developed. Here, a short introduction to $\in_T$-logics is given, which can express impredicative quantification, total truth predicates, and self-reference.

1 Introduction

The elementary concepts of set formation can be regarded as primitives in the construction of extensions. This perspective to set-theory takes the view of sets as models and of set-theory as a theory of "definability". The foundational nature of extensions made set-theory the classical and most widely used ontological basis and proof theory of mathematics. A large part of the mathematical language consists of notions that have sets as their meaning. In particular, if we speak of numbers, pairs, relations, functions, sequences, families, graphs, paths, equivalences, equations, algebras, structures, topologies, spaces and the like we refer, in each case, to particularly defined sets, classes or constructs. Generally, we consider a mathematical notion to be "well-defined", if its meaning is a properly defined set. Well-definedness in this sense gives identity to the notion that is given its set-theoretic meaning. Unrestricted set-comprehension, however, may lead to contradictions as it is the case in the "set" of all sets that do not contain themselves as members. In set-theory the problem of unrestricted formation of sets is therefore resolved by a proper collection of set-theoretic axioms. Conventionally among those axioms are the axioms of extensionality and of foundation, which state that two sets are equal if and only if they have the same extensions and that the membership relation within a set does not allow an infinite descend. The notion of function, as it is established in the well-known λ-calculus, makes the basis of functional programming in computer science but can not be given a set-theoretic meaning if the conventional set-theoretic axioms are assumed to hold. For example, self-applicable functions, which appear naturally in computer science, can be expressed in the λ-calculus, but as sets fail to obey the axiom of foundation. It has long been an open question if there are models of the λ-calculus at all. It was only Dana Scott's domain-theoretic model which answered this question by the construction of infinite limits. But it is still today true that the set-theoretic notion of function and the applicative notion of function, as it is modelled in the λ-calculus, do not coincide.

In the following we introduce a new style of semantics for set-like modelling by interpreting ϵ-formulas and sentences in their conventional first-order semantics, thereby allowing unrestricted forms of membership relations. We will see that the conventional reading of such models as sets is not affected.

In particular, we introduce ϵ-structures as the semantic domains for the interpretation of set-like concepts which we express in terms of ε-logic formulas. Using ϵ-sets and ε-logic the conventional set-theoretic concepts of pairs, relations, and functions are being simulated. We then show that reflexive domains with self-applicable functions can naturally be defined and used to build a model of the λ-calculus in the sense of Hindley and Seldin (1986). These investigations show that the claim of set-like modelling without the axioms of extensionality and foundation can be fulfilled. ϵ-structures give rise to the notion of ϵ-sets which may be not well-founded and may be intensional, as they need not to be equal if their extensions are the same. The idea of ϵ-sets was developed independentely from Aczel's theory of non-wellfounded sets, which replaces the axiom of foundation by the antifoundation axiom, (see Aczel (1988) or Bairwise and Moss (1996), for example). To study the relationship between the two it is necessary to discuss factorizations which enforce extensionality. A comparison yields the fact that Aczel sets correspond to ϵ-sets after factorization of the ϵ-sets by a maximal extension conform bisimulation. We conclude the discussion of ϵ-style semantics with a brief introduction to $\in_T$-logics which were originally motivated from the concepts of ϵ-sets. $\in_T$-logics are propositional logics with identity, quantification over propositional variables and the truth predicates ": true" and ": false". $\in_T$-logics fulfill the Tarksi biconditionals in that their (non-partial) semantics is consistent with the interpretation of their truth predicates. The relationship between $\in_T$-logics and ϵ-structures is established by the fact that semantic truth- and falseness is not modelled by symbols "true" and "false", but rather by sets of propositions which belong to the ϵ-set *true* and the ϵ-set *false*, respectively.

The approach to set-like modelling proposed here originated from observations on untyped semantics for hierarchically typed algebraic specifications, dating back to the late eighties. It grew out of intensive discussions about increased expressiveness of formal descriptions mainly for natural language semantics and processing. The results of these discussions have first been published in the technical report of Mahr et al. (1990). ϵ-theory, as we phrased this approach, was later studied in view of typetheoretic applications (see Mahr (1993)) and was applied in several ways, namely by Kutsche (1994), Pooyan (1992) and Pooyan-Weihs (1999), Glas (1997), Umbach (1996), Such (1994), and Tunjic (2008); a philosophical reflexion of ϵ-theory by Robering is found in the introduction of the collection Robering (1994). $\in_T$-logics, first introduced by Sträter (1992a), were further studied by Zeitz (2000) and are presently under investigation in regard to logic integration, see Bab and Mahr (2005) and extensions by modalities, see Bab (2005) and Bab (2007). ϵ-sets were first investigated by Sträter (1992b). After a period of sleep ϵ-theory was picked up again in 2006 in student seminars and is now presented here in this paper.

2 Basic Definitions

In this section the basic definitions to develop ϵ-style semantics are given. We use set-theoretic notions and work in ZFC, the Zermelo-Frankel set theory together with the axiom of choice. This is done model-theoretically in a "Tarski-style": We assume that a universe of sets V together with the membership-relation $\in$ on V is given, such that $(V, \in)$ follows the ZFC-axioms. Henceforth we assume that all objects we use belong to V. Within this frame, ϵ-style semantics is developed. Thus ZFC may be seen as the theory underlying ϵ-style semantics.

For the axiomatic system ZFC, results in ZFC and consistency results we refer to Kunen (1980) and Jech (2002).

2.1 ϵ-Structures

We start with the definitions of ϵ- and set-structures.

Definition 1 (ϵ-structure). *An ϵ-structure is a tupel $\mathcal{M} = (M, \epsilon_M)$ where M is a nonempty set of objects (or entities) and ϵ_M is a binary relation on M.* ⊟

An important class of ϵ-structures are *set-structures*:

Definition 2 (Set-structure).

1. *A set-structure is an ϵ-structure $\mathcal{M} = (M, \epsilon_M)$ with $\epsilon_M = \in_M$, where $\in_M$ is the membership relation restricted to M, i.e., $\in_M = \{(a, b) \in M \times M : a \in b\}$. For every set S in ZFC the set-structure $(S, \in_S)$ is denoted by $\mathcal{M}_S$ and is called the set-structure generated by S.*
2. *A set-structure $\mathcal{M}_S$ is transitive if S is a transitive set, i.e., if, for every $a \in S$, if $b \in a$ then $b \in S$. For every set S in ZFC we call $\mathcal{M}_{\text{tc}(S)}$*

the transitive set-structure generated by S, where $\operatorname{tc}(S)$ is the transitive closure of S.[1]

Hence two ϵ-structures are associated with a ZFC-set S, the set-structure $\mathcal{M}_S$ and the transitive set-structure $\mathcal{M}_{\operatorname{tc}(S)}$. Note that $S = \operatorname{tc}(S)$ iff S is transitive. See the following example[2]:

Example 1 (Set-structure). Let $S = \{1, 3\}$, where natural numbers are defined as usual by $0 := \emptyset$ and $n+1 := \{0, \ldots, n\}$. Then $\operatorname{tc}(S) = \{0, 1, 2, 3\} = 4$ and

$$\mathcal{M}_S = (\{1, 3\}, \in_S), \text{ where } \in_S = \{(1, 3)\} \text{ and}$$

$$\mathcal{M}_{\operatorname{tc}(S)} = (4, \in_4), \quad \text{where } \in_4 = \{(0, 1), (0, 2), (0, 3), (1, 2), (1, 3), (2, 3)\}.$$

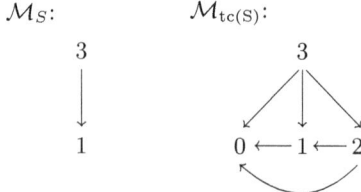

Fig. 1.

Moreover, $\operatorname{tc}(\operatorname{tc}(S)) = \operatorname{tc}(S) = 4$, so $\mathcal{M}_4 = \mathcal{M}_{\operatorname{tc}(4)}$.

Other examples of ϵ-structures are the following:

Example 2 (Pair). Let A, C_1, C_2, a, b be arbitrary sets and define the ϵ-structure $\mathcal{M}_{\text{pair}} = (M_{\text{pair}}, \epsilon_{pair})$ by

$M_{\text{pair}} = \{A, C_1, C_2, a, b\}$ and $\epsilon_{\text{pair}} = \{(C_1, A), (C_2, A), (a, C_1), (a, C_2), (b, C_2)\}$.

It will be shown later that this is the canonical ϵ-structure to model a *pair*. It is a set-structure if and only if $A = \{C_1, C_2\}$, $C_1 = \{a\}$, $C_2 = \{a, b\}$.

---

[1] The transitive closure of a set S, $\operatorname{tc}(S)$, is the minimal superset of S, which is transitive. It is inductively defined by $\bigcup^0 S = S$, $\bigcup^{n+1} S = \bigcup(\bigcup^n S)$ and $\operatorname{tc}(S) := \bigcup\{\bigcup^n S : n \in \mathbb{N}\}$.

[2] We give visualizations of ϵ-structures $\mathcal{M} = (M, \epsilon_M)$ or of parts of them in the form of graphs, where $A \to B$ means that $B \epsilon_M A$. If different points within a visualization are denoted by the same object, they represent identical entities in M.

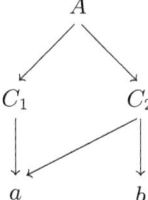

Fig. 2. $\mathcal{M}_{\text{pair}}$

Example 3 (Loop). $\mathcal{M}_{\text{loop}} = (\{\Omega\}, \{(\Omega, \Omega)\})$ is an ϵ-structure. It is known as the *loop* of hyperset theory. □

Example 4 (Infinite descending chain). $\mathcal{M}_{\text{idc}} = (M_{\text{idc}}, \epsilon_{\text{idc}})$ with $M_{\text{idc}} = \mathbb{N}$ and $\epsilon_{\text{idc}} = \{(n+1, n) : n \in \mathbb{N}\}$ is an ϵ-structure which models an *infinite descending chain*. □

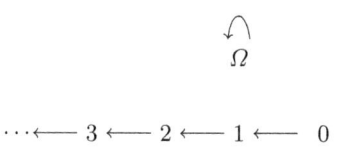

Fig. 3. $\mathcal{M}_{\text{loop}}$ and $\mathcal{M}_{\text{idc}}$

Note that Examples 3 and 4 fail to be set-structures because the ϵ-relation in both cases is not well-founded (see Theorem 1 at the end of the section).

2.2 ε-logic

ϵ-structures may be seen as first order structures with only one binary relation ϵ. To speak formally about properties of ϵ-structures and of elements of it, we introduce first order ϵ-logic whose conventional interpretation are ϵ-structures.

Definition 3 (ε-formulas). *Let X be an infinite set of variables, denoted by $x, y, z, u, v, w, x_1, x_2, \ldots$. The set L_ε of ε-formulas is recursively defined as follows:*

1. *$(x = y)$ and $(x\varepsilon y)$ are the atomic formulas in L_ε for every $x, y \in X$.*
2. *If $\varphi, \psi \in L_\varepsilon$ then so are $(\neg\varphi)$ and $(\varphi \wedge \psi)$.*
3. *If $\varphi \in L_\varepsilon$ and $x \in X$ then $(\forall x.\varphi) \in L_\varepsilon$.*

Hence L_ε are the first order formulas of predicate calculus over X with equality and the single binary relation-symbol ε. □

We will use the symbols $\vee, \rightarrow, \leftrightarrow$ and $\exists$ in the usual way to abbreviate complex formulas. For example, $(\varphi \rightarrow \psi)$ abbreviates $((\neg\varphi) \vee \psi)$ and $(\exists x.\varphi)$ stands for $(\neg(\forall x.(\neg\varphi)))$. Binding conventions are used as usual to drop unnecessary parentheses. For further notations and concepts see also Ehrig et al. (2001).

Note 1. There are three different symbols for "epsilon" that are to be distinguished: The first one is ϵ which is used to denote the binary relation of ϵ-structures. The second one is $\in$ which stands for membership of sets in V. The third one is ε which is being used in formulas, see Definition 3 above. □

ϵ-structures are sets with a binary relation on it and thus are the classical structures to interpret ε-formulas: Let $\mathcal{M} = (M, \epsilon_M)$ be some ϵ-structure, $\varphi \in L_\varepsilon$ and $\beta : X \rightarrow M$ be an assignment which gives every variable a value in M. Then we ask whether or not $(\mathcal{M}, \beta) \models \varphi$, where "$\models$" is the usual satisfaction relation and ε is interpreted as ϵ_M. The coincidence lemma, which holds for ε-logic, allows for a simplified notation of satisfaction, which we will use in the following for convenience: If the free variables of φ are $x_1 \ldots x_n$, where $n \geq 0$, we write $\varphi(x_1, \ldots, x_n)$ for φ, and if $\beta(x_1) = a_1, \ldots \beta(x_n) = a_n$, we write
$$\mathcal{M} \models \varphi[a_1, \ldots, a_n] \quad \text{instead of} \quad (\mathcal{M}, \beta) \models \varphi.$$
The following are examples of ε-formulas with free variables:

$$\varphi_{\text{union}}(x, y, z) := \forall u.(u\varepsilon x \leftrightarrow u\varepsilon y \vee u\varepsilon z) \tag{1}$$

$$\varphi_{\text{intersection}}(x, y, z) := \forall u.(u\varepsilon x \leftrightarrow u\varepsilon y \wedge u\varepsilon z) \tag{2}$$

$$\varphi_{\text{subset}}(x, y) := \forall z.(z\varepsilon x \rightarrow z\varepsilon y) \tag{3}$$

$$\varphi_{\text{successor}}(x) := \exists y.(x\varepsilon y \wedge \forall z.(z\varepsilon x \rightarrow z\varepsilon y)) \tag{4}$$

$$\varphi_{\text{found}}(x) := \exists y.(y\varepsilon x) \rightarrow \exists u.(u\varepsilon x \wedge \neg\exists z.(z\varepsilon x \wedge z\varepsilon u)) \tag{5}$$

Existential quantification in $\varphi_{\text{found}}(x)$ is of different flavour than in $\varphi_{\text{successor}}(x)$. In $\varphi_{\text{found}}(x)$ quantification has the form $\exists y.(y\varepsilon x \wedge \psi)$, so y *is bounded by the free variable* x. Bounded quantification is best motivated by using $\wedge$, $\neg$ and $\exists$ as a base for L_ε. In the following formulas of the form $\exists x\varepsilon y.\varphi$ abbreviate $\exists x.(x\varepsilon y \wedge \varphi)$. Formulas with only bounded quantifiers are usually called Δ_0-formulas (see Kunen (1980) or Jech (2002)).

Definition 4 (Δ_0-formulas). *The set $\Delta_0 \subseteq L_\varepsilon$ is recursively defined as follows:*

1. *$x = y$ and $x\varepsilon y$ are in Δ_0, if x, y are variables in X.*
2. *If φ, ψ are in Δ_0 then so are $\neg\varphi$ and $\varphi \wedge \psi$.*
3. *If φ is in Δ_0 and $x, y \in X$ with $x \neq y$, then $\exists x\varepsilon y.\varphi$ is in Δ_0.*

If $\varphi \in L_\varepsilon$ is logically equivalent to some $\psi \in \Delta_0$,[3] we say that φ is Δ_0.

---

[3] Two formulas $\varphi(x_1, \ldots, x_n)$ and $\psi(y_1, \ldots, y_m)$ are called *logically equivalent* if, for every ϵ-structure $\mathcal{M} = (M, \epsilon_M)$ and every assignment $\beta : X \rightarrow M$, $(\mathcal{M}, \beta) \models \varphi(x_1 \ldots, x_n)$ iff $(\mathcal{M}, \beta) \models \psi(y_1, \ldots, y_m)$.

Note that formulas of the form $\forall x \varepsilon y . \varphi$ are Δ_0, too, if φ is Δ_0: The bounded universal quantifier is an abbreviation for $\forall x . (x \varepsilon y \to \varphi)$ and this abbreviates $\neg \exists x \in y . \neg \varphi$. Hence $\forall x \varepsilon y . \varphi$ is Δ_0 if φ is. Consequently $\varphi_{\text{subset}}(x, y)$, $\varphi_{\text{union}}(x, y, z)$, $\varphi_{\text{intersection}}(x, y, z)$, and $\varphi_{\text{found}}(x)$ are Δ_0. $\varphi_{\text{successor}}(x)$ fails to be Δ_0. Δ_0-formulas have an important property which concern validity within substructures, see Lemma 2 below.

The following are examples of ε-sentences, i.e., ε-formulas without free variables:

$$\psi_{\text{ext}} := \forall x . \forall y . (\forall z . (z \varepsilon x \leftrightarrow z \varepsilon y) \to x = y) \tag{6}$$

$$\psi_{\text{found}} := \forall x . (\exists y . (y \varepsilon x) \to \exists u . (u \varepsilon x \wedge \neg \exists z . (z \varepsilon x \wedge z \varepsilon u))) \tag{7}$$

$$\psi_{\text{Russel}} := \exists x . \forall y . (y \varepsilon x \leftrightarrow \neg y \varepsilon y) \tag{8}$$

$$\psi_{\text{loop}} := \exists x . x \varepsilon x \tag{9}$$

$$\psi_{\text{idc}} := \forall y . \exists x . x \varepsilon y \tag{10}$$

The first two sentences are the restricting axioms of ZFC, the *axiom of extensionality* (ψ_{ext}) and the *axiom of foundation* (ψ_{found}). The first one is valid in transitive set-structures and the latter in every set-structure, see Proposition 1 below. ψ_{Russel} describes *Russel's Paradox* and is a contradiction, i.e., it is invalid in every ϵ-structure. ψ_{loop} and ψ_{idc} can have a meaning in ϵ-structures, see $\mathcal{M}_{\text{loop}}$ and $\mathcal{M}_{\text{idc}}$, but have no meaning in set-structures.

The following facts are well known (see Kunen (1980), Chapter IV, for example)

Proposition 1. *Let $\mathcal{M}$ be an ϵ-structure.*

1. *If $\mathcal{M}$ is a set-structure then $\mathcal{M} \models \psi_{\text{found}}$.*
2. *If $\mathcal{M}$ is a transitive set-structure then $\mathcal{M} \models \psi_{\text{ext}}$.* ⊟

Note 2. In both cases the converse is not true: For example, $\mathcal{M}_{\text{idc}} \models \psi_{\text{found}}$ and $\mathcal{M}_{\text{idc}} \models \psi_{\text{ext}}$, but $\mathcal{M}_{\text{idc}}$ is not a set-structure. Moreover, if one considers a set $S = \{a, b\}$ with $a \neq b$, $a \notin b$ and $b \notin a$, then $\mathcal{M}_S$ is a non-transitive set structure in which φ_{ext} fails. ⊟

Remark 1 (Summary). We recalled that ϵ-structures are the classical models to interpret ε-formulas. Gödel's Completeness-Theorem implies that it is possible to find for every consistent $\Phi \subseteq L_\varepsilon$ an ϵ-structure $\mathcal{M}$ with $\mathcal{M} \models \Phi$. To the contrary, not for every consistent $\Phi \subseteq L_\varepsilon$ there is a set-structure $\mathcal{M}_S$ with $\mathcal{M}_S \models \Phi$, of which ψ_{loop} and ψ_{idc} are examples. Therefore, if classical sets are being identified with their generated set-structures, it turns out that ϵ-structures provide a more general semantic interpretation than sets which are restricted by the axioms of foundation.[4] ⊟

---

[4] For a presentation of Gödel's Completeness-and Incompleteness-Theorems we refer to Rautenberg (1996).

2.3 Constructions: ϵ-substructures and ϵ-homomorphisms

If two objects $a, b \in M$ of an ϵ-structure $\mathcal{M} = (M, \epsilon_M)$ are given, the fact that
$$b \epsilon_M a$$
is read as a statement of membership. For a given $a \in M$ we are interested in the ϵ-*members* of a.

Definition 5 (ϵ-extensions). *Let $\mathcal{M} = (M, \epsilon_M)$ be an ϵ-structure and let $a \in M$. Then*
$$\mathrm{ext}_\mathcal{M}(a) := \{b \in M : b \epsilon_M a\}$$
is the ϵ-extension of a in $\mathcal{M}$. If $b \in \mathrm{ext}_\mathcal{M}(a)$, b is an ϵ-member of a in $\mathcal{M}$. We call $\mathcal{M}$ extensional if for all $a, b \in M$
$$\mathrm{ext}_\mathcal{M}(a) = \mathrm{ext}_\mathcal{M}(b) \text{ implies } a = b.$$

The following is obvious:

Lemma 1. *Let $\mathcal{M}$ be an ϵ-structure. Then $\mathcal{M}$ is extensional iff $\mathcal{M}$ satisfies the axiom of extensionality, i.e., iff $\mathcal{M} \models \psi_{\mathrm{ext}}$.*

By Proposition 1 every transitive set-structure $\mathcal{M}_S$ is extensional, and the extension of an element $a \in S$ is a itself. For this reason, in the theory of independence proofs for set theory one normally works with transitive models (or even elementary submodels of transitive models), in order to enforce φ_{ext} to hold: in ZFC, sets *are their extensions*. In contrast to that, in arbitrary ϵ-structures or even non-transitive set-structures there may be different objects with the same ϵ-extension. So: in the case of ϵ-structures, *sets have an ϵ-extension*.

The interest to consider transitive models as models of set theory motivates the definition of ϵ-substructures. Classically, a substructure of a relational structure (M, R) in model theory simply is a structure (N, R') with $N \subseteq M$ and $R' = R \upharpoonright (N \times N)$. But with this concept of substructure, an entity of the substructure may loose some of its ϵ-members. ϵ-substructures will therefore have to obey the stronger assumption that extensions are being preserved:

Definition 6 (ϵ-substructures). *Let $\mathcal{M} = (M, \epsilon_M)$ and $\mathcal{N} = (N, \epsilon_N)$ be ϵ-structures. Then $\mathcal{N}$ is an ϵ-substructure of $\mathcal{M}$ and $\mathcal{M}$ is an ϵ-superstructure of $\mathcal{N}$ if the following hold:*

1. $N \subseteq M$;
2. *If $a \in N$ then $\mathrm{ext}_\mathcal{N}(a) = \mathrm{ext}_\mathcal{M}(a)$.*

Note this implies that $\epsilon_N = \epsilon_M \restriction (N \times N)$. Hence an ϵ-substructure of $\mathcal{M}$ is a substructure which is *transitive with respect to* ϵ_M. This transitivity guarantees an important feature concerning the satisfaction of formulas: obviously, if an arbitrary ε-formula φ holds in some ϵ-structure this does not indicate that it holds in every ϵ-sub- or superstructure of it. For example, if an $\exists$-quantifier occurs in φ then φ may fail to hold within an ϵ-substructure, and an $\forall$-quantifier may be satisfied by an ϵ-substructure but not by an ϵ-superstructure. But in ϵ-substructures satisfaction is preserved if the formulas have only *bounded* quantifiers. This is stated in the following lemma:

Lemma 2. *Let $\mathcal{M} = (M, \epsilon_M)$ be an ϵ-structure and let $\mathcal{N} = (N, \epsilon_N)$ be some substructure of $\mathcal{M}$. Let $\varphi(x_1, \ldots, x_n)$ be Δ_0 and let $a_1 \ldots, a_n \in N$. Then*

$$\mathcal{M} \models \varphi[a_1, \ldots, a_n] \quad \Leftrightarrow \quad \mathcal{N} \models \varphi[a_1, \ldots, a_n].$$

□

In the theory of independence proofs one says in this case that φ is *absolute for* $\mathcal{N}$, $\mathcal{M}$. The proof of the lemma is essentially the same as the proof of the fact that a Δ_0-formula holds in a transitive set-structure iff it is true in the universe (see Jech (2002, 165f.)). The lemma indicates, for example, that the formula $\varphi_{\text{subset}}(x, y)$ is true in an ϵ-substructure iff it is true in its ϵ-superstructure, i.e., if $\mathcal{N} = (N, \epsilon_N)$ is an ϵ-substructure of $\mathcal{M}$ and $A, B \in N$ then

$$\mathcal{M} \models \varphi_{\text{subset}}[A, B] \quad \text{iff} \quad \mathcal{N} \models \varphi_{\text{subset}}[A, B].$$

For every ϵ-structure $\mathcal{M} = (M, \epsilon_M)$ and for every $A \subseteq M$ we define the minimal ϵ-substructure of $\mathcal{M}$ containing A as follows:

Definition 7 (generated ϵ-substructures). *Let $\mathcal{M} = (M, \epsilon_M)$ be an ϵ-structure and $A \subseteq M$ with $A \neq \emptyset$.*

1. *We define $M(A)$ recursively by*
   a) $A \subseteq M(A)$;
   b) if $b \in M(A)$ then $\text{ext}_\mathcal{M}(b) \subseteq M(A)$.
   Then $\mathcal{M}(A) = (M(A), \epsilon_M \restriction (M(A) \times M(A)))$ is the ϵ-substructure of $\mathcal{M}$ generated by A.
2. *If $A = \{a_1, \ldots, a_n\}$ is finite we write $\mathcal{M}(a_1, \ldots, a_n)$ instead of $\mathcal{M}(\{a_1, \ldots, a_n\})$.*

□

The notion of ϵ-homomorphisms allows to relate and compare ϵ-structures.

Definition 8 (ϵ-homomorphism). *Let $\mathcal{M} = (M, \epsilon_M)$ and $\mathcal{N} = (N, \epsilon_N)$ be ϵ-structures and $h : M \to N$ be a mapping.*

1. *h is an ϵ-homomorphism if for all $a \in M$*

$$\text{ext}_\mathcal{N}(h(a)) = \{h(b) : b \in \text{ext}_\mathcal{M}(a)\}.$$

2. *h is an ϵ-monomorphism or an ϵ-embedding if it is one-to-one.*
3. *h is an ϵ-epimorphism if it is surjective onto N. In this case we call $\mathcal{M}$ and $\mathcal{N}$ ϵ-homomorphic.*
4. *h is an ϵ-isomorphism if it is an ϵ-homomorphism and bijective. In this case we call $\mathcal{M}$ and $\mathcal{N}$ ϵ-isomorphic.* □

While ϵ-isomorphisms coincide with the classical notion of isomorphism (see Lemma 3 below) ϵ-homomorphisms are more special. Even the notion of *strong homomorphism* (see Rautenberg (1996), for example), which is normally used in model theory, does not fit the intention, that images of homomorphisms are ϵ-substructures. Therefore the definition of an ϵ-homomorphism is chosen to be stronger.

Lemma 3. *Let $h : M \to N$ be an ϵ-homomorphism between ϵ-structures $\mathcal{M} = (M, \epsilon_M)$ and $\mathcal{N} = (N, \epsilon_N)$. Then*

1. *$(h[M], \epsilon_N \upharpoonright (h[M] \times h[M]))$ is an ϵ-substructure of $\mathcal{N}$. We call it the ϵ-image of h.*
2. *h is an ϵ-isomorphism if and only if h is bijective and*

$$a \epsilon_M b \quad \Leftrightarrow \quad h(a) \epsilon_N h(b)$$

for all $a, b \in M$.
3. *Let $g : N \to N'$ be another ϵ-homomorphism between $\mathcal{N}$ and some ϵ-structure $\mathcal{N}' = (N', \epsilon_{N'})$. Then $g \circ h$ is an ϵ-homomorphism between $\mathcal{M}$ and $\mathcal{N}'$.* □

We obviously have the following:

Lemma 4. *Let $h : M \to M'$ be an ϵ-monomorphism between ϵ-structures $\mathcal{M} = (M, \epsilon_M)$ and $\mathcal{M}' = (M', \epsilon_{M'})$. Let $\mathcal{N} = (N, \epsilon_N)$ be the ϵ-image of h, $\varphi(x_1, \ldots, x_n) \in L_\varepsilon$ and $a_1, \ldots, a_n \in M$. Then*

$$\mathcal{M} \models \varphi[a_1, \ldots, a_n] \quad \textit{iff} \quad \mathcal{N} \models \varphi[h(a_1), \ldots, h(a_n)].$$

□

The one-to-one correspondence between $\mathcal{M}$ and the ϵ-image of h is the crucial factor for the preservation of validity. By applying this lemma to ϵ-isomorphisms, it turns out that, using Proposition 1, an ϵ-structure does not have an isomorphic set-structure if it fails to obey the axiom of foundation, ψ_{found}. It has already been pointed out that the converse is not true, because $\mathcal{M}_{\text{idc}} \models \psi_{\text{found}}$. We close this section by characterizing those ϵ-structures for which an isomorphic set-structure exists.

Definition 9 (Well-founded relations). *A relation $R \subseteq A \times A$ on a set A is called* well-founded *if, for every nonempty subset $X \subseteq A$, there exists some $a \in X$ such that $(b, a) \notin R$ for any $b \in X$. In this case a is called R-minimal in X.* □

Without proof we state the following

Theorem 1. *Let $\mathcal{M} = (M, \epsilon_M)$ be an ϵ-structure. Then the following are equivalent:*

1. *ϵ_M is a well-founded relation on M.*
2. *There is a set S such that $\mathcal{M}_S$ is isomorphic to $\mathcal{M}$.* ⊟

3 ϵ-theoretic concepts and their ϵ-style semantics

In this section we show how one can model in ϵ-structures and with ϵ-structures in a set-like style. For this purpose the notion of an ϵ-theoretic concept and its meaning is developed. As an application of ϵ-style modelling we show how one can build an ϵ-structure which is a model of the λ-calculus. First, we introduce ϵ-sets, which are ϵ-structures generated by a single object.

3.1 ϵ-sets

The notion of ϵ-set is motivated by the fact that members of ZFC-sets are sets again. It uses the fact that objects of ϵ-structures can in a natural way be identified with ϵ-sets. Recall that, for a given ϵ-structure $\mathcal{M} = (M, \epsilon_M)$ and some $a \in M$, $\mathcal{M}(a)$ is the minimal substructure of $\mathcal{M}$ generated by a, see Definition 7 above.

Definition 10 (ϵ-sets). *Let $\mathcal{M} = (M, \epsilon_M)$ be an ϵ-structure. $\mathcal{M}$ is an ϵ-set if $\mathcal{M} = \mathcal{M}(a)$ for some $a \in M$. We say that $\mathcal{M}$ is minimal pointed to a.* ⊟

A nice characterization of being an ϵ-set is the following. It is easily proven by induction:

Lemma 5. *Let $\mathcal{M} = (M, \epsilon_M)$ be an ϵ-structure and $a \in M$. Then the following are equivalent:*

1. *$\mathcal{M}$ is an ϵ-set minimal pointed to a;*
2. *For every $b \in M$ there are finitely many $a_0 \ldots a_n$ with*

$$b = a_n \epsilon a_{n-1} \epsilon \ldots \epsilon a_0 = a. >$$

⊟

Obviously not every ϵ-structure is an ϵ-set and set-structures may fail to be ϵ-sets, too. But for a given ϵ-structure $\mathcal{M} = (M, \epsilon_M)$ we can set $M' := M \cup \{M\}$ and $\epsilon_{M'} := \epsilon_M \cup \{(a, M) : a \in M\}$. Then $\mathcal{M}' = (M', \epsilon_{M'})$ is an ϵ-set. This assignment is one-to-one and strictly monotone in the sense that $\mathcal{M}$ is a proper ϵ-substructure of $\mathcal{M}'$.

If $\mathcal{M}$ is an ϵ-set which is minimal pointed to a, then a is not necessarily unique in having this property, because there may be some $b \in \mathcal{M}(a)$ with $\mathcal{M} = \mathcal{M}(a) = \mathcal{M}(b)$. This is, for example, the case if $a\epsilon b$ and $b\epsilon a$.

If we use an ϵ-structure $\mathcal{M} = (M, \epsilon_M)$ as a semantic model, we treat every object $a \in M$ as if it were identical with $\mathcal{M}(a)$, the minimal ϵ-substructure of $\mathcal{M}$ which contains a as a member. This is justified by the fact that in the modelling with ϵ-formulas, variables which reference sets are being treated alike.

Note that ϵ-sets within a transitive set-structure $\mathcal{M}_S$ do not coincide with the ZFC-sets which are the members of S. Namely, if $T \in S$ then T is identified with the ϵ-set

$$\mathcal{M}_S(T) = \mathcal{M}_{\{T\} \cup T}$$

and not with $\mathcal{M}_T$. Note also that $\mathcal{M}_S(T)$ is again a transitive set-structure.

3.2 ϵ-theoretic Concepts

Semantically, formulas with free variables describe properties of entities within a given structure. Consider as an example

$$\varphi_{\text{singleton}}(x) = \exists y.(y\varepsilon x \wedge \forall z.(z\varepsilon x \to y = z))$$

and let $\mathcal{M} = (M, \epsilon_M)$ and $a \in M$. Then $\mathcal{M} \models \varphi_{\text{singleton}}[a]$ if and only if a is a singleton in $\mathcal{M}$, i.e., if there is $b \in M$ with $\text{ext}_{\mathcal{M}}(a) = \{b\}$.

An example of a formula with more than one free variable is

$$\varphi_{\text{pair}}(x, y, z) = \exists u_1 \varepsilon x. \exists u_2 \varepsilon x. (\ y\varepsilon u_1 \wedge y\varepsilon u_2 \wedge z\varepsilon u_2 \wedge \forall v\varepsilon x.(v = u_1 \vee v = u_2)$$
$$\wedge \forall w \varepsilon u_1.w = y \wedge \forall w\varepsilon u_2.(w = y \vee w = z)).$$

Then $\mathcal{M}_{\text{pair}} \models \varphi_{\text{pair}}[A, a, b]$ where $\mathcal{M}_{\text{pair}}$ is defined as in Example 2. So A is the Kuratowski-pair of a and b in $\mathcal{M}_{\text{pair}}$.

We call $\varphi_{\text{singleton}}(x)$ the *concept of a singleton* and $\varphi_{\text{pair}}(x, y, z)$ the *concept of a Kuratowski-pair of given components*. In general:

Definition 11 (ϵ-theoretic concept). Let $\varphi(x_1, \ldots, x_n) \in L_\varepsilon$, $n \geq 1$.

1. $\varphi(x_1, \ldots, x_n)$ is called an (n-ary) ϵ-theoretic concept.
2. Let $\mathcal{M} = (M, \epsilon_M)$ be an ϵ-structure and $a_1, \ldots, a_n \in M$. We call $(\mathcal{M}, a_1, \ldots, a_n)$ an instance of $\varphi(x_1, \ldots, x_n)$ if
   a) $\mathcal{M} \models \varphi[a_1, \ldots, a_n]$ and
   b) $\mathcal{M}$ is minimal with respect to property a), i.e., there is no proper ϵ-substructure $\mathcal{N} = (N, \epsilon_N)$ of $\mathcal{M}$ with $a_1, \ldots, a_n \in N$ and $\mathcal{N} \models \varphi[a_1, \ldots, a_n]$.
3. The class of all instances of $\varphi(x_1, \ldots, x_n)$ is called the meaning of $\varphi(x_1, \ldots, x_n)$ and is denoted by $\mu\langle\varphi(x_1, \ldots, x_n)\rangle$. □

The notion of ϵ-theoretic concept is meant to formally state the idea of *set formation*, however in the broader frame of ϵ-structures and the modelling of ϵ-sets.

Using Lemma 2, the meaning of an ϵ-theoretic concept is easily characterized if $\varphi(x_1, \ldots, x_n)$ is Δ_0:

Proposition 2. Let $\varphi(x_1,\ldots,x_n)$ be an ϵ-theoretic concept which is Δ_0. Let $\mathcal{M} = (M, \epsilon_M)$ be an ϵ-structure and $a_1, \ldots, a_n \in M$. Then the following are equivalent:

1. $(\mathcal{M}, a_1, \ldots, a_n) \in \mu\langle\varphi(x_1,\ldots,x_n)\rangle$;
2. $\mathcal{M} \models \varphi[a_1,\ldots,a_n]$ and $\mathcal{M} = \mathcal{M}(a_1,\ldots,a_n)$. □

Corollary 1. Let $\varphi(x)$ be a Δ_0-concept with only one free variable. Then $(\mathcal{M}, a) \in \mu\langle\varphi(x)\rangle$ implies that $\mathcal{M}$ is an ϵ-set minimal pointed to a, i.e., that $\mathcal{M}(a) = \mathcal{M}$. □

Example 5 (The meaning of $\varphi_{\text{pair}}(x,y,z)$). Consider again $\varphi_{\text{pair}}(x,y,z)$ which is Δ_0. By the previous proposition, for every $\mathcal{M} = (M, \epsilon_M)$ and $A, a, b \in M$, if $\mathcal{M} \models \varphi_{\text{pair}}[A, a, b]$, then $(\mathcal{M}(A, a, b), A, a, b) \in \mu\langle\varphi_{\text{pair}}(x,y,z)\rangle$. Thus, for every ZFC-pair (a,b), there are two canonical instances of $\varphi_{\text{pair}}(x,y,z)$: $(\mathcal{M}_{\{(a,b),\{a\},\{a,b\},a,b\}}, (a,b), a, b)$ and $(\mathcal{M}_{\text{tc}(\{(a,b)\})}, (a,b), a, b)$.

By the minimality of instances one could be tempted to think that, if $(\mathcal{M}, A, a, b)$ belongs to the meaning of $\varphi_{\text{pair}}(x,y,z)$, then the object A is *the* ordered pair of a and b in $\mathcal{M}$. But this may not be justified. Let for example

$$\mathcal{M} = (\{a, b, C_1, C_2\}, \{(a, C_1)(a, C_2), (b, C_2), (C_1, a), (C_2, a), (C_1, b), (C_2, b)\})$$

for different a and b. Then $(\mathcal{M}, a, a, b)$ and $(\mathcal{M}, b, a, b)$ are instances of $\varphi_{\text{pair}}(x,y,z)$, i.e., both lie in $\mu\langle\varphi_{\text{pair}}(x,y,z)\rangle$, because a and b are both pairs with components a and b in $\mathcal{M}$. Obviously, such $\mathcal{M}$ cannot be a set-structure because ϵ_M is not well-founded. □

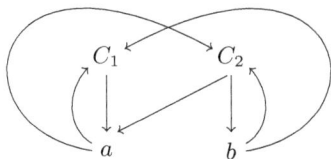

Fig. 4. a and b are both ordered pairs of a and b

The characterization of $\mu\langle\varphi(x_1,\ldots,x_n)\rangle$ for Δ_0-concepts fails for general concepts: Consider for example $\varphi_{\text{successor}}(x) = \exists y.(x\varepsilon y \wedge \forall z.(z\varepsilon x \to z\varepsilon y))$ which is not Δ_0. Then for the transitive set-structure $\mathcal{M}_4$ (see Example 1) $(\mathcal{M}_4, 2) \in \mu\langle\varphi_{\text{successor}}(x)\rangle$ because 3 witnesses the existence of a successor for 2 and is minimal in the sense of condition 2b of Definition 11, but $\mathcal{M}_4$ is not minimal pointed to 2.

3.3 Semantic Modelling: ϵ-pairs, ϵ-relations, and ϵ-functions

We now turn to ϵ-concepts of particular interest. They have one free variable and are being used in set theory to denote *pairs*, *relations* and *functions*. The concept of a pair of given components, $\varphi_{\text{pair}}(x, y, z)$, was already discussed in Example 5 and it was stressed that the meaning of $\varphi_{\text{pair}}(x, y, z)$, which defines Kuratowski-pairs out of given components provides a much wider class of instances than only the classical ZFC-pairs and their corresponding set-structures.

Consider the concept $\varphi_{\text{ispair}}(x)$ with only one free variable x:

$$\varphi_{\text{ispair}}(x) = \exists u_1 \varepsilon x. \exists u_2 \varepsilon x. \exists y. \varepsilon u_1. \exists z. \varepsilon u_2. (y \varepsilon u_2 \wedge \psi),$$

where ψ is the tail of $\varphi_{\text{pair}}(x, y, z)$. $\varphi_{\text{ispair}}(x)$ is Δ_0, so that by Corollary 1 the ϵ-structures of the instances of φ_{ispair} are ϵ-sets. Again, the meaning of $\varphi_{\text{ispair}}(x)$ contains more instances than the classical Kuratowski-pairs of ZFC, like, for example, $(\mathcal{M}_{\text{loop}}, \Omega) \in \mu\langle\varphi_{\text{ispair}}(x)\rangle$. We call every ϵ-set $\mathcal{M} = (M, \epsilon_M)$ with $(\mathcal{M}, a) \in \mu\langle\varphi_{\text{ispair}}(x)\rangle$ an ϵ-*pair with respect to* a. The conceptualization of an ϵ-pair shows well the modelling principle discussed in 3.1, and already applied in 3.2: the formulas which define the conventional set theoretic notions are being read as ϵ-theoretic concepts with their broader meaning.

For a further demonstration of this principle consider the following two concepts: First, the ϵ-concept of a *relation*, which means for a set in ZFC a *set whose members are Kuratowski-pairs*:

$$\varphi_{\text{rel}}(x) := \forall z \varepsilon x. \varphi_{\text{ispair}}(z).$$

Second, the ϵ-concept of a *function*, which means for a ZFC-set a *relation which is right unique*:

$$\varphi_{\text{fun}}(x) := \varphi_{\text{rel}}(x) \wedge \forall z_1 \varepsilon x. \forall z_2 \varepsilon x. (\forall y_1 \varepsilon z_1. \exists x_1 \varepsilon y_1. \forall y_2 \varepsilon z_2. x_1 \varepsilon y_2 \to z_1 = z_2).$$

Both concepts are Δ_0. We therefore have by Corollary 1 that, if $(\mathcal{M}, a)$ belongs to the meaning of $\varphi_{\text{rel}}(x)$ or $\varphi_{\text{fun}}(x)$, then $\mathcal{M}$ is an ϵ-set with respect to a. This also motivates the following definition:

Definition 12 (ϵ-pairs, ϵ-relations, ϵ-functions). *Let $\mathcal{M} = (M, \epsilon_M)$ be an ϵ-structure and $a \in M$. Then we say*

1. *if $(\mathcal{M}(a), a)$ is an instance of $\varphi_{\text{ispair}}(x)$ then a is a pair in $\mathcal{M}$ and $\mathcal{M}(a)$ is an ϵ-pair with respect to a;*
2. *if $(\mathcal{M}(a), a)$ is an instance of $\varphi_{\text{rel}}(x)$ then a is a relation in $\mathcal{M}$ and $\mathcal{M}(a)$ is an ϵ-relation with respect to a;*
3. *if $(\mathcal{M}(a), a)$ is an instance of $\varphi_{\text{fun}}(x)$ then a is a function in $\mathcal{M}$ and $\mathcal{M}(a)$ is an ϵ-function with respect to a.*

From Lemma 2 we conclude

Corollary 2. Let $\mathcal{M} = (M, \epsilon_M)$ be an ϵ-structure and $a \in M$.

1. a is a pair in $\mathcal{M}$ iff $\mathcal{M} \models \varphi_{\text{pair}}[a]$.
2. a is a relation in $\mathcal{M}$ iff $\mathcal{M} \models \varphi_{\text{rel}}[a]$.
3. a is a function in $\mathcal{M}$ iff $\mathcal{M} \models \varphi_{\text{fun}}[a]$. $\square$

Within transitive set-structures the notions of being an ϵ-pair (an ϵ-relation or an ϵ-function resp.) and of being a pair (relation, function) in the usual set-theoretic sense, do coincide:

Lemma 6. Let $\mathcal{M} = (M, \epsilon_M)$ be a transitive set-structure and $a \in M$.

1. a is a pair in ZFC iff a is a pair in $\mathcal{M}$ iff $\mathcal{M}(a)$ is an ϵ-pair.
2. a is a relation in ZFC iff a is a relation in $\mathcal{M}$ iff $\mathcal{M}(a)$ is an ϵ-relation.
3. a is a function in ZFC iff a is a function in $\mathcal{M}$ iff $\mathcal{M}(a)$ is an ϵ-function. $\square$

Lemma 6 explains in which sense we can speak of ϵ-sets as a conservative extension of sets.

Note 3. Contrary to ZFC-sets an ϵ-set $\mathcal{M}$ can be both, an ϵ-function and an ϵ-pair and there are two different situations in which this happens: First, if there are different objects a and b to which $\mathcal{M}$ is minimal pointed. So, if we say that $\mathcal{M}$ is an ϵ-function (ϵ-pair, ϵ-relation, resp.) we should always add "with respect to", because we can look at an ϵ-set from *different points of view*.[5]

Second, $\mathcal{M}_{\text{loop}}$ (see Example 3 above) is both, an ϵ-function and an ϵ-pair with respect to the same entity Ω, so $(\mathcal{M}_{\text{loop}}, \Omega)$ is an instance of $\psi_{\text{ispair}}(x)$ and of $\psi_{\text{fun}}(x)$. But this is, up to isomorphism, the only ϵ-structure with this property. $\square$

In ZFC there is no function f which can map itself or is itself included in the range (*i.e.*, $(f, a) \notin f$ and $(a, f) \notin f$ for any a). But there are ϵ-functions with that property and the following example shows how one can model ϵ-functions which are given by their operational behaviour.[6]

Example 6. Let a, f_1, f_2 and f_3 be pairwise different sets. We define an ϵ-set $\mathcal{M} = (M, \epsilon_M)$ as it is visualized in Figure 6.

Then $\mathcal{M}$ is minimal pointed to a such that f_1, f_2, f_3 are functions in $\mathcal{M}$ with the operational behaviour $f_1(f_i) = f_i$ and $f_2(f_i) = f_i$ for $i = 1, 2, 3$ and $f_3(f_1) = f_2$, $f_3(f_2) = f_3$ and $f_3(f_3) = f_1$. The objects which are ordered pairs are written as (f_1, f_1) for example. So $\mathcal{M} = \mathcal{M}(a)$ is an ϵ-set which models three functions which contain themselves as components.

---

[5] If we say that $\mathcal{M}(a)$ is an ϵ-pair (ϵ-relation, ϵ-function resp.) we always imply that it is one with respect to a.

[6] The example is taken from Glas (1997).

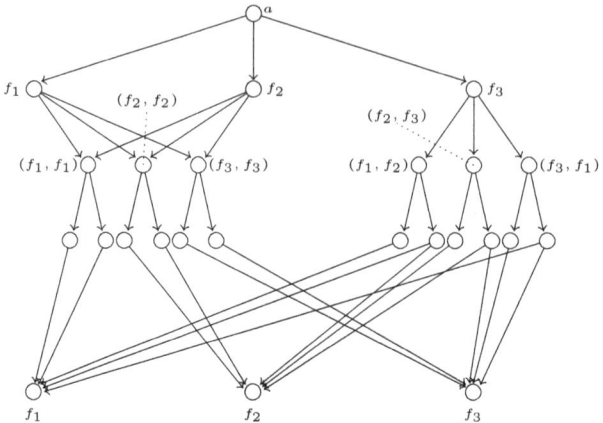

Fig. 5. Example 6: A functional-reflexive ϵ-set

We have $\mathcal{M}(f_1) = \mathcal{M}(\langle f_1, f_1 \rangle)$, so this is an ϵ-set which is an ϵ-function with respect to one entity and an ϵ-pair with respect to another entity and is therefore an example for the situation discussed as first case in Note 3.

$\epsilon_\mathcal{M}$ obviously is not well-founded on $\mathcal{M}$ and as well it is not on $\mathcal{M}(f_i)$. So, by Theorem 1, there are no set-structures isomorphic to $\mathcal{M}$ or $\mathcal{M}(f_i)$. ⊟

The example shows that with ϵ-sets one can quite directly model functions defined by operational descriptions without getting in conflict with restricting set-theoretic axioms. ϵ-sets, like $\mathcal{M}(a)$ in the previous example, are called *functional-reflexive*.

Definition 13 (Functional-reflexive ϵ-sets). *Let $\mathcal{M}$ be some ϵ-set minimal pointed to a. $\mathcal{M}(a)$ is called* functional reflexive with respect to a *if, for every $f \epsilon a$, the following hold:*

1. *$\mathcal{M}(f)$ is an ϵ-function with domain $\text{ext}_\mathcal{M}(a)$, and*
2. *if $p \epsilon f$ is a pair of b and c in $\mathcal{M}$ then $b \epsilon a$ and $c \epsilon a$.* ⊟

There is a Δ_0-formula $\varphi_\text{refl}(x)$ which describes both conditions of the definition in such a way that an ϵ-set $\mathcal{M}(a)$ is reflexive if and only if $\mathcal{M}(a) \models \varphi_\text{refl}[a]$ iff $(\mathcal{M}(a), a)$ is an instance of $\varphi_\text{refl}(x)$. Accordingly, as for the notions of pairs, relations and functions, for an arbitrary entity a within some ϵ-structure $\mathcal{M}$, $\mathcal{M}(a)$ is functional-reflexive iff $\mathcal{M} \models \varphi_\text{refl}[a]$, and we say in that case that a is *functional-reflexive in $\mathcal{M}$*.

3.4 Semantic modelling: An ϵ-model for λ-calculus

It will now be shown how ϵ-models for the λ-calculus can be built with functional-reflexive ϵ-sets. This has been done by Pooyan (1992) for the classical pure λ-calculus and by Glas (1997) for an advanced λ-calculus which respects applicative order reduction. We follow Glas (1997) in our construction of an ϵ-model for the pure λ-calculus. For the λ-calculus and its models we refer to the book of Hindley and Seldin (1986).

The λ-calculus is known as a theory for formalizing functions and function application. It is used in computer science as a base of functional programming. The set of λ-terms $\Gamma(V)$ over a set of variables $V := \{u, v, w, u_1, \dots\}$ is defined as follows:

$$\begin{array}{lll} \text{variable} & u \in \Gamma(V) & \text{if } u \in V; \\ \text{application} & (EF) \in \Gamma(V) & \text{if } E, F \in \Gamma(V); \\ \text{abstraction} & (\lambda u.E) \in \Gamma(V) & \text{if } u \in V \text{ and } E \in \Gamma(V). \end{array}$$

Parentheses can be omitted with the convention, that they are associated to the left. The "$\lambda u.$" binds the variable u in the succeeding expression. Since the concepts of bounded and free variables, renaming and substitutions are defined in the usual way we do not give formal definitions, assuming the reader's familiarity with these concepts. α-congruence $=_\alpha$ is a congruence-relation on λ-terms which identifies two λ-terms E and F, written $E =_\alpha F$, iff E and F syntactically differ at most in the names of their bounded variables in such a way that no name-clash occurs. The substitution $F[u := E]$ of a variable u by a λ-term E in F is defined by replacing every free occurrence of u by E such that the free variables of E remain free in $F[u := E]$. This might require to rename bounded variables of F.

A λ-term E may be interpreted as a function, and a term EF as the application of E to the λ-term F. If $E = \lambda u.E'$ is an abstraction, the result of the application is $E'[u := F]$. This process of function application is formulated by the β-rule

$$(\beta) \frac{(\lambda u.E)F}{E[u := F]}.$$

The β-congruence-relation $=_\beta$ on λ-terms is the least congruence-relation which extends $=_\alpha$ and identifies each β-redex with its *contractum*. $=_\beta$ describes application but it is intensional, *i.e.*, there are λ-terms E and F such that $EG =_\beta FG$ for every $G \in \Gamma(V)$ but $E \neq_\beta F$. To enforce extensionality of the functions defined by the calculus, the rule η is used:

$$(\eta) \frac{\lambda u.Eu}{E} \text{ if } u \text{ is not free in } E$$

$=_{\beta\eta}$ is the least congruence relation closed under the β and η-rule, and implies that $E =_{\beta\eta} F$ if and only if $EG =_{\beta\eta} FG$ for all $G \in \Gamma(V)$.

A model for the λ-calculus is a mathematical object in which every λ-term can be mapped such that β or $\beta\eta$ congruences are respected and functional application can be expressed within the model. Naively, one would like to take set theoretic functions as a domain for the model, but classical set theoretic functions do not allow self application (in ZFC). To define a λ-model one builds an *applicative structure* as its domain which is a set D together with a binary operation $\cdot : D \times D \to D$, in a way that $(D, \cdot)$ becomes a λ-model. Then there are two different ways to design D, one uses the syntax of λ-terms to declare the applicative structure, and the other is syntax free in the sense that it does not use the notion of λ-terms. The two ways, however, are related in the sense that for a given syntax-free λ-model with domain $(D, \cdot)$ one can construct a λ-model which is not syntax free, but has the same domain, and vice versa. In the following we discuss models which are not syntax-free.

In Hindley and Seldin (1986), the definition of a non-syntax free λ-model works as follows: A λ-model for β is an applicative structure $(D, \cdot)$, together with a mapping $[\![\,]\!] : \Gamma(V) \times {}^V D \to D$, $[\![\,]\!](E, \varrho)$ written as $[\![E]\!]_\varrho$, such that the following properties hold:

(1) $\quad [\![u]\!]_\varrho = \varrho(u)$
(2) $\quad [\![EF]\!]_\varrho = [\![E]\!]_\varrho \cdot [\![F]\!]_\varrho$
(3) $[\![\lambda u.E]\!]_\varrho \cdot d = [\![E]\!]_{\varrho[u:=d]}$
(4) $\quad [\![E]\!]_\varrho = [\![E]\!]_\sigma$ if $\varrho(u) = \sigma(u)$ for all $u \in \text{free}(E)$
(5) $\quad [\![\lambda u.E]\!]_\varrho = [\![\lambda v.E[u := v]]\!]_\varrho$ if $v \notin \text{free}(E)$
(6) $\quad [\![\lambda u.E]\!]_\varrho = [\![\lambda u.F]\!]$ if $[\![E]\!]_{\varrho[u:=d]} = [\![F]\!]_{\varrho[u:=d]}$ for all $d \in D$.

If additionally

(7) $\quad [\![\lambda u.Eu]\!]_\varrho = [\![E]\!]$ for every u not free in E

is satisfied, one speaks of a model for $\beta\eta$ and calls the resulting model *extensional*. Clearly these conditions should hold if one intends the λ-model to respect $=_{\beta\eta}$ and in Hindley and Seldin (1986) it is shown that they imply that every equation provable in $=_{\beta\eta}$ is true in the λ-model, *i.e.*, if $E =_{\beta\eta} F$ then $[\![E]\!]_\varrho = [\![F]\!]_\varrho$ for every $\varrho : V \to D$.

Now let $\mathcal{M} = \mathcal{M}(D)$ be a functional-reflexive ϵ-set minimal pointed to D. Then every $f \in \text{ext}_\mathcal{M}(D)$ is an ϵ-function in $\mathcal{M}(D)$ with domain and range $\text{ext}_\mathcal{M}(D)$. Thus, for $g \epsilon D$ there is a unique ϵ-pair in $\text{ext}(f)$ which has g as its first component. We denote the second component of this unique pair by $f\langle g\rangle$. It lies in $\text{ext}_\mathcal{M}(D)$ again, so if we define $\circ_\mathcal{M} : \text{ext}_\mathcal{M}(D) \times \text{ext}_\mathcal{M}(D) \to \text{ext}_\mathcal{M}(D)$ by $f \circ_\mathcal{M} g = f\langle g\rangle$, $(\text{ext}_\mathcal{M}(D), \circ_\mathcal{M})$ becomes an applicative structure. This motivates the following definition of an ϵ-λ-model:

Definition 14 (ϵ-λ-model). *Let $\mathcal{M}_D$ be a functional-reflexive ϵ-set minimal pointed to D and let $[\![\,]\!] : \Gamma(V) \times V \to \text{ext}(D)$. $(\mathcal{M}_D, [\![\,]\!])$ is an ϵ-λ-model for β if the applicative structure $(\text{ext}(D), \circ_\mathcal{M})$ together with $[\![\,]\!]$ is a λ-model, i.e., if the requirements (1)-(6) above are satisfied. If additionally (7) holds, $(\mathcal{M}_D, [\![\,]\!])$ is an ϵ-λ-model for $\beta\eta$.* ⌐

So every ϵ-λ-model provides a syntactical λ-model in the sense of Hindley and Seldin, and it is not difficult to define a reflexive ϵ-set out of an applicative structure which is an ϵ-λ-model for β ($\beta\eta$ resp.) if the applicative structure is a λ-model for β ($\beta\eta$ resp.). The main difference between these two notions of model is that in the case of ϵ-λ-models λ-terms can properly be interpreted as functions, namely as ϵ-*functions*, despite the fact that the definition of ϵ-λ-models is syntax dependent. Hence, following our modelling principle of ϵ-semantics, if f is the interpretation of a λ-term F in an ϵ-λ-model, f reflects the operational behaviour of F with the ϵ-relation of the reflexive ϵ-set in a set-like style. We will demonstrate this by showing how an ϵ-λ-model can be constructed:

Let $\Gamma(V)_{\beta\eta}$ be the set of equivalence-classes of $\Gamma(V)$ with respect to $=_{\beta\eta}$. For every $E \in \Gamma(V)$ we denote the equivalence-class by $[E]$, and define an ϵ-structure $\mathcal{M}^\lambda = (M^\lambda, \epsilon_\lambda)$ by:

$$M^\lambda := \{\Gamma(V)_{\beta\eta}\} \cup \Gamma(V)_{\beta\eta} \cup (\Gamma(V)_{\beta\eta} \times \Gamma(V)_{\beta\eta}) \cup \{\{[E],[F]\} : E, F \in \Gamma(V)\}$$

and ϵ_λ by defining the extensions of the objects in M^λ:

$$\mathrm{ext}_\lambda(\Gamma(V)_{\beta\eta}) := \Gamma(V)_{\beta\eta}$$
$$\mathrm{ext}_\lambda([E]) := \{([F],[EF]) : F \in \Gamma(V)\}$$
$$\mathrm{ext}_\lambda(([E],[F])) := \{\{[E]\},\{[E],[F]\}\}$$
$$\mathrm{ext}_\lambda(\{[E],[F]\}) := \{[E],[F]\}.$$

Then $\mathcal{M}^\lambda$ is an ϵ-set minimally pointed to $\Gamma(V)_{\beta\eta}$. The only crucial point is to see that $([F],[G])$ lies in $\mathrm{ext}_\lambda([E])$ for some λ-term E: consider any $E := \lambda x.G$ where x does not occur in G. Then $EF = (\lambda x.G)F =_\beta G$, so $([F],[G]) \in \mathrm{ext}_\lambda([E])$. Therefore $\mathcal{M}^\lambda = (\mathcal{M}^\lambda)(\Gamma(V)_{\beta\eta})$. This is a functional-reflexive ϵ-set pointed to $\Gamma(V)_{\beta\eta}$, and within this ϵ-set we have

1. for every $[E] \in \Gamma(V)_{\beta\eta}$ we have $[E]\epsilon_\lambda\Gamma(V)_{\beta\eta}$, and
2. every $[E]$ is an ϵ-function which maps $[F]$ to $[EF]$.

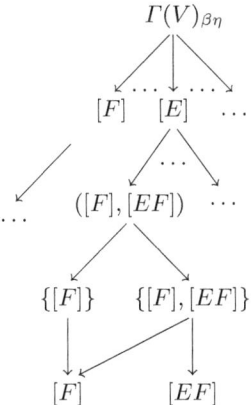

Fig. 6. ϵ-λ-Term-Model

Now we set, for every $\varrho : V \to \Gamma(V)_{\beta\eta}$,

$$[\![E]\!]^\lambda_\varrho = [E[u_1 := \varrho(u_1),\ldots, u_n := \varrho(u_n)]] \text{ if free}(E) = \{u_1,\ldots, u_n\}.$$

With this definition conditions (1)–(7) are being satisfied. We conclude

Proposition 3 (Term-Model). $(\mathcal{M}^\lambda, [\![]\!]^\lambda)$ *is an ϵ-λ-model for $\beta\eta$.* □

Note 4. By Definition 14 of an ϵ-λ-model and by the discussion following it, every syntactical λ-model in the sense of Hindley and Seldin can be translated into an ϵ-λ-model and vice versa. A slightly different way to obtain a λ-model from an ϵ-λ-model is as follows: Take the ϵ-functions of the functional-reflexive set of an ϵ-λ-model as the domain and define $\circ$ to be the application. Then a syntactical λ-model results, whose entities *are* set-theoretic functions within the frame of ϵ-style semantics. Hence, ϵ-λ-models provide a new type of syntactical λ-models in the sense of Hindley and Seldin. For more results on ϵ-λ-models and syntax-free forms of them, see Tunjic (2008). □

4 Extensionality

ϵ-structures do not have to be extensional but always have extensional homomorphic images. In this section two constructions of extensional homomorphic images of a given ϵ-structure are presented.

4.1 ϵ-quotients

Every homomorphic image of an ϵ-structure is isomorphic to a certain quotient of the ϵ-structure. So we first introduce the notion of ϵ-quotients.

Definition 15 (ϵ-quotients). *Let $\mathcal{M} = (M, \epsilon_M)$ be an ϵ-structure and $\sim$ be some equivalence-relation on M. Set*

$$M/\sim := \{[a]_\sim : a \in M\} \text{ and}$$

$$\epsilon_\sim := \{([a]_\sim, [b]_\sim) : \text{there are } a', b' \in M \text{ with } a' \sim a,\ b' \sim b \text{ and } a'\epsilon b'\}.$$

Then $\mathcal{M}/\sim := (M/\sim, \epsilon_\sim)$ is an ϵ-structure, the ϵ-quotient of $\mathcal{M}$ with respect to $\sim$. □

For a given ϵ-structure $\mathcal{M}$ the extension conform equivalence relations produce ϵ-quotients which are homomorphic to $\mathcal{M}$:

Definition 16 (extension-conformequivalence). *Let $\mathcal{M} = (M, \epsilon_M)$ be some ϵ-structure and let $\sim$ be an equivalence relation on M. Then $\sim$ is extension conform if for all $A, A' \in M$ the following holds:*

$$A \sim A' \quad \Rightarrow \quad \text{for all } a\epsilon A \text{ there exists } a'\epsilon A' \text{ with } a \sim a'.$$

□

The following characterization can easily be proven from Definitions 15 and 16.

Lemma 7. *Let $\mathcal{M} = (M, \epsilon_M)$ be an ϵ-structure and let $\sim$ be an equivalence relation on M. Then the following are equivalent:*

1. *$\sim$ is extension conform.*
2. *The canonical projection $\pi : M \to M/\sim$, defined by $\pi(a) = [a]$, is an ϵ-epimorphism from $\mathcal{M}$ onto $\mathcal{M}/\sim$, i.e., $\text{ext}([a]) = \{[b] : b \in \text{ext}_\mathcal{M}(a)\}$.* ⊟

Quotients are known in algebra for various objects. Their definitions are justified by the well known *Factorization Theorems*. For ϵ-structures an according theorem can be stated as follows:

Proposition 4 (Factorization Theorem). *Let $h : M \to N$ be some ϵ-homomorphism between ϵ-structures $\mathcal{M} = (M, \epsilon_M)$ and $\mathcal{N} = (N, \epsilon_N)$. Define $\sim_h \subseteq M \times M$ by*

$$a \sim_h b \quad \Leftrightarrow \quad h(a) = h(b).$$

Then $\sim_h$ is an extension conform equivalence relation. Moreover, let π be the canonical projection from $\mathcal{M}$ onto $\mathcal{M}/\sim_h$ and define $i : M/\sim_h \to N$ by $i([a]) = h(a)$. Then i is an ϵ-monomorphism and $i \circ \pi = h$. ⊟

The proof is straight forward using the Definition 8 of ϵ-homomorphism and Lemma 7 above.

Consider the special case that h is an ϵ-epimorphism. Then i is onto and one-to-one, in which case $\mathcal{N}$ is isomorphic to $\mathcal{M}/\sim$:

Corollary 3. *Let $h : M \to N$ be some ϵ-epimorphism mapping $\mathcal{M} = (M, \epsilon_M)$ onto $\mathcal{N} = (N, \epsilon_N)$. Then $\mathcal{N}$ is isomorphic to an ϵ-quotient of $\mathcal{M}$.* ⊟

We close this subsection by stating the fact that quotients of ϵ-sets are ϵ-sets again.

Lemma 8. *Let $\mathcal{M}$ be some ϵ-set and let $\sim$ be an extension conform equivalence relation on $\mathcal{M}$. Then $\mathcal{M}/\sim$ is an ϵ-set.* ⊟

4.2 Extensional ϵ-quotients

In the following we show how extensional quotients of a given ϵ-structure $\mathcal{M}$ can be build. Let $\mathcal{M} = (M, \epsilon_M)$ be an ϵ-structure. The main idea of how to turn $\mathcal{M}$ into an extensional ϵ-structure is to identify objects in M which have identical extensions, i.e., to factorize $\mathcal{M}$ with the equivalence relation $\sim_{\text{ext}}$, defined by $a \sim_{\text{ext}} b$ iff $\text{ext}_\mathcal{M}(a) = \text{ext}_\mathcal{M}(b)$. Obviously $\sim_{\text{ext}}$ is extension conform but in general $\mathcal{M}/\sim_{\text{ext}}$ fails to be extensional, as the following example shows:

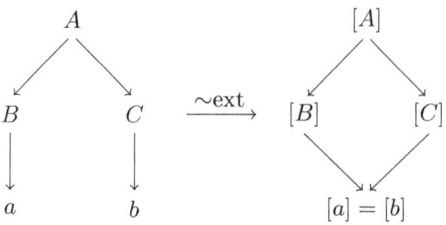

Fig. 7. $\mathcal{M}/\sim_{\text{ext}}$ is not extensional

To end up with an extensional structure we therefore iterate this procedure of identifying objects of $\mathcal{M}_1 := \mathcal{M}/\sim_{\text{ext}}$ with equal extensions and get $\mathcal{M}_2 := \mathcal{M}_1/\sim_{\text{ext}}$ and, inductively, $\mathcal{M}_{i+1} := \mathcal{M}_i/\sim_{\text{ext}}$. By the Factorization Theorem and Lemma 3.3, $\mathcal{M}_i$ is a homomorphic image of $\mathcal{M}$ which may be considered to be a quotient of $\mathcal{M}$ with respect to an extension conform equivalence relation $\sim_{\text{ext}}^i$ on M. But again $\mathcal{M}_i$ may fail to be extensional for all $i \in \mathbb{N}$, so that we have to consider limit cases. Formally this will be done by introducing an operator $()^+$ which is applied to binary relations on ϵ-structures and extends extension conform equivalence relations to new ones. This operator is an adaptation of Aczel's operator $()^+$ for relations on systems.

Definition 17 (($()^+$). Let $\mathcal{M} = (M, \epsilon_M)$ be an ϵ-structure and let $R \subseteq M \times M$. Then $R^+ \subseteq M \times M$ is defined by

$$(a,b) \in R^+ \quad \Leftrightarrow \quad \text{for all } c\epsilon a \text{ there is some } d\epsilon b \text{ with } cRd \text{ and}$$
$$\text{for all } d\epsilon b \text{ there is some } c\epsilon a \text{ with } dRc.$$

□

The proofs of the following lemma and proposition can be found in Glas (1997).

Lemma 9. Let $\mathcal{M} = (M, \epsilon_M)$ be an ϵ-structure.

1. The operator $()^+$ is monotone, i.e., if R_1 and R_2 are binary relations on M with $R_1 \subseteq R_2$, then $R_1^+ \subseteq R_2^+$.
2. Let $\sim$ be an extension conform equivalence relation on $\mathcal{M}$. Then
   a) $\sim \subseteq \sim^+$.
   b) $\sim^+$ is an extension conform equivalence relation.

□

Let $\mathcal{M} = (M, \epsilon_M)$ be an ϵ-structure and consider $\sim_{\text{ext}}$ on $\mathcal{M}$. If $\mathcal{M}$ is already extensional, then $\sim_{\text{ext}}$ is nothing else but the minimal equivalence relation $\sim_{\min} = \{(a,a) : a \in M\}$, and $\mathcal{M}/\sim_{\text{ext}}$ is isomorphic to $\mathcal{M}$. Moreover, $\sim_{\text{ext}}$ is the result of $()^+$ applied to $\sim_{\min}$, so $\sim_{\min}$ is a fixpoint of $()^+$ if $\mathcal{M}$ is extensional. For arbitrary extension conform equivalences we have

Proposition 5. *Let $\mathcal{M}$ be an ϵ-structure and $\sim$ be an extension conform equivalence relation on $\mathcal{M}$. Then $\sim = \sim^+$ iff $\mathcal{M}/\sim$ is extensional.* ▯

We now define the smallest extension conform equivalence relation for which the quotient is extensional. "Smallest" means that this equivalence is a subset of every other extension conform equivalence relation which provides an extensional quotient. So it is secured that this equivalence relation only identifies those objects which have to be identified to make the ϵ-structure extensional.

Let $\mathcal{M} = (M, \epsilon_M)$ be an ϵ-structure. We define a chain $(\sim_\beta)_{\beta \in \mathrm{On}}$[7] inductively by

$$\sim_0 \; := \; \sim_{\min}$$
$$\sim_{\alpha+1} \; := \; \sim_\alpha^+, \text{ and}$$
$$\sim_\delta \; := \; \bigcup_{\alpha < \delta} \sim_\alpha \text{ if } \delta \text{ is a limit ordinal.}$$

Then, using Lemma 9, an easy induction shows that $\sim_\beta$ is an extension conform equivalence relation for every $\beta \in \mathrm{On}$.

Setting

$$\sim_{\mathrm{Ext}} \; := \; \bigcup_{\beta \in \mathrm{On}} \sim_\beta$$

one can show that this is the smallest extension conform equivalence relation on $\mathcal{M}$ such that $\sim_{\mathrm{Ext}}^+ = \sim_{\mathrm{Ext}}$: Extension conformness is shown by induction, using Lemma 9, again. The monotonicity of $()^+$ implies that $\sim_{\mathrm{Ext}} \subseteq \sim_{\mathrm{Ext}}^+$ (see Lemma 9.2), and equality follows from cofinality arguments. If $\sim$ is any other extension conform equivalence relation on $\mathcal{M}$ with $\sim^+ = \sim$ then by an easy induction one concludes that $\sim_\beta \subseteq \sim$ for every $\beta \in \mathrm{On}$, starting with the trivial fact that $\sim_{\min} = \sim_0 \subseteq \sim$. Thus $\sim_{\mathrm{Ext}} \subseteq \sim$ and we have:

Proposition 6. *Let $\mathcal{M} = (M, \epsilon_M)$ be some ϵ-structure and $\sim_{\mathrm{Ext}}$ be defined as above. Then $\sim_{\mathrm{Ext}}$ is the smallest extension conform equivalence relation such that $\mathcal{M}/\sim_{\mathrm{Ext}}$ is extensional.* ▯

Note 5. One can do the same construction by starting with an arbitrary extension conform equivalence relation $\sim$ instead of $\sim_{\min}$. This would lead to the minimal extension conform equivalence relation which extends $\sim$ and makes the quotient extensional. ▯

We now turn to another extreme by identifying objects of an ϵ-structure as most as possible by an extension conform equivalence relation. This equivalence relation is the union of all extension conform equivalence relations of the given ϵ-structure. The quotient defined by this equivalence is a homomorphic image of every other quotient:

---

[7] On is the class of all ordinals.

Proposition 7. *Let $\mathcal{M} = (M, \epsilon_M)$ be an ϵ-structure and set*

$$\sim_{\max} := \bigcup \{\sim : \sim \text{ is an extension conform equivalence relation}\}.$$

Then:

1. *$\sim_{\max}$ is the maximal extension conform equivalence relation on $\mathcal{M}$ and $\mathcal{M}/\sim_{\max}$ is extensional.*
2. *Let $\mathcal{N}$ be a homomorphic image of $\mathcal{M}$. Then there is a unique ϵ-epimorphism from $\mathcal{N}$ to $\mathcal{M}/\sim_{\max}$. We say that $\mathcal{M}/\sim_{\max}$ is terminal in the class of all ϵ-structures which are homomorphic to $\mathcal{N}$.* □

For a proof we refer to Glas (1997).

5 ϵ-sets and Aczel's Theory

We now compare ϵ-style semantics with Aczel's theory of non well-founded sets. First a very rough overview of Aczel's antifoundation axiom AFA and one possible model of ZFC$^-$+AFA is given. Then ϵ-sets are compared with accessible pointed graphs and it is described how ϵ-sets can be mapped into Aczel's universe. We close the section by discussing the main differences of ϵ-semantics and the theory of Aczel.

The main sources of this theory are Aczel (1988) and, for example, the book of Bairwise and Moss (1996). Details may be looked up there. Our overview is taken from the appendix of the Set-Theory book of Moschovakis (1994), which also gives a short introduction to Aczel's theory.

5.1 Aczel's Antifoundation-Axiom AFA and a Model of ZFC$^-$+AFA

A *graph* is a pair $(G, \to_G)$ of a set G and a binary relation $\to_G \subseteq G \times G$. An *accessible pointed graph (apg)* is a triple $\mathcal{G} = (G, \to_G, p_G)$ such that $(G, \to_G)$ is a graph, $p_G \in G$, and with the property that, for every $a \in G$, there is a finite sequence $a_0, \ldots, a_n$ with $p_G = a_0, \to_G a_1 \to_G \cdots \to_G a_n = a$. If $(G, \to_G)$ is a graph and $p \in G$ then $\mathcal{G}_p = (G_p, \to_G, p)$ is the *least subgraph generated by p* which is an apg pointed to p, i.e., $a \in G_p$ if there is a finite sequence $a_0, \ldots a_n$ with $p = a_0 \to_G \cdots \to_G a_n = a$. A graph $(G, \to_G)$ (an apg $\mathcal{G} = (G, \to_G, p_G)$ resp.) is *well-founded* if there is no infinite sequence $(a_i)_{i \in \mathbb{N}}$ with $a_i \in G$ and $a_i \to_G a_{i+1}$.

A *decoration* of a graph $(G, \to_G)$ is a map d which assigns to every $p \in G$ some set $d(p)$ such that $d(p) = \{d(q) : p \to_G q\}$. An apg $(G, \to_G, p_G)$ is a *picture of a set A* if there is a decoration d of $(G, \to_G)$ such that $d(p_G) = A$. One version of Mostowski's Collapsing Theorem which can be proved in ZFC$^-$ is the following:

Theorem 2 (Mostowski's Collapsing Theorem). *Every well-founded graph has a unique decoration.* □

It follows that every well-founded apg is the picture of a unique set and this set is well-founded. Moreover, still in ZFC$^-$, it is easy to see, that every set A has a picture: Just take the *transitive closure* TC(A) of A as the domain of the graph and the reverse element relation as edges. Then $(TC(A), \in^{-1}, A)$ is an apg and the identity map $d(a) = a$ for $a \in TC(A)$ is a decoration with $d(A) = A$. But obviously in ZFC, an apg which is not well-founded does not have a decoration because it would result in a non well-founded set. Hence Aczel's anti-foundation axiom is:

AFA: *Every graph has a unique decoration.*

AFA states that non well-founded graphs are pictures of sets, too, and these sets cannot be well-founded. Moreover, AFA has a restricting part by saying that an apg can only be the picture of one unique set.

In order to sketch how one can build a model of ZFC$^-$+AFA out of a model of ZFC, we introduce the notion of bisimulation: A relation $R \subseteq G \times H$ between two accessible pointed graphs $(G, \to_G, p_G)$ and $(H, \to_H, p_H)$ is a *bisimulation* if $p_G R p_H$ and R satisfies the following implication:

$$aRb \Rightarrow \text{for all } c \to_G a \text{ there is some } d \to_H b \text{ with } cRd \text{ and}$$
$$\text{for all } d \to_H b \text{ there is some } c \to_G a \text{ with } dRc.$$

Two apg's $\mathcal{G}$ and $\mathcal{H}$ are *bisimular* if there exists a bisimulation between them. We write $\mathcal{G} =_{\text{bs}} \mathcal{H}$. Then $=_{\text{bs}}$ is an equivalence relation on the class of all apg's. For well-founded apg's we have the following fact: $\mathcal{G} =_{\text{bs}} \mathcal{H}$ if and only if they are pictures of the same set. Assuming AFA, this is true for non well-founded graphs, too. The main idea of the *Antifounded Universe* is to identify graphs which are bisimular. The universe may be constructed as follows.

Let $\mathcal{V}$ be a model of ZFC. Consider the following subclass of all pointed graphs in $\mathcal{V}$:

$$\mathcal{A}_0 := \{\mathcal{G} \in \mathcal{V} : \mathcal{G} = (G, \to_G, p_G) \text{ is an apg}\}$$

together with the binary relation ϵ_0

$$\mathcal{G} \epsilon_0 \mathcal{H} :\Leftrightarrow \text{ there is some } q \in H \text{ with } \mathcal{G} =_{\text{bs}} \mathcal{H}_q$$

for every $\mathcal{G}, \mathcal{H} \in \mathcal{A}_0$. We now identify the apg's of $\mathcal{A}_0$ which are bisimular by taking the quotient

$$\mathcal{A} := \mathcal{A}_0/=_{\text{bs}} \quad \text{and define} \quad [\mathcal{G}]_{\text{bs}} \epsilon_{\text{afa}} [\mathcal{H}]_{\text{bs}} :\Leftrightarrow \mathcal{G} \epsilon_0 \mathcal{H}.$$

This definition is independent of the choice of representatives, because ϵ_0 respects bisimulations, *i.e.*, if $\mathcal{G}_1 \epsilon_0 \mathcal{H}_1$, $\mathcal{G}_1 =_{\text{bs}} \mathcal{G}_2$, and $\mathcal{H}_1 =_{\text{bs}} \mathcal{H}_2$ then $\mathcal{G}_2 \epsilon_0 \mathcal{H}_2$.[8]

---

[8] We should note that our definition of $\mathcal{A}$ is not quite exact, because for a given apg $\mathcal{G} \in \mathcal{V}$ the collection of all apg's which are bisimular to $\mathcal{G}$ is a proper class,

Theorem 3. $(\mathcal{A}, \epsilon_{\text{afa}})$ *is a model for* ZFC$^-$+AFA. □

Thus we have the following relative consistency result, that, if a model for ZFC exists, then there exists a model of ZFC$^-$+AFA.[9] By the fact, that one can build a model of ZFC out of a model of ZFC$^-$, it follows that a model of ZFC$^-$+AFA exists iff a model for ZFC$^-$ exists.

We now compare ϵ-sets with Aczel-sets.

5.2 Mapping ϵ-sets into Aczel's Universe

Every graph may be considered as an ϵ-structure by taking the inverse relation $\leftarrow_G := \rightarrow_G^{-1}$ as ϵ and vice versa. If $\mathcal{M} = (M, \epsilon_M)$ is an ϵ-set minimal pointed to a then, by the characterization of ϵ-sets (see Lemma 5) $\mathcal{G}(\mathcal{M})_a := (M, \epsilon_M^{-1}, a)$ is an apg. Conversely, for every apg $(G, \rightarrow_G, p_G)$ the associated ϵ-structure is minimal pointed to p_G.

Moreover the notion of bisimulation is closely related to extension-conform equivalence relations. Let $\mathcal{M} = (M, \epsilon_M)$ be an ϵ-set minimal pointed to a, i.e., $\mathcal{M} = \mathcal{M}(a)$, and let $\sim$ be an extension conform equivalence relation on $\mathcal{M}$. Then by Lemma 8 $\mathcal{M}/\sim$ is an ϵ-set minimal pointed to $[a]_\sim$. Define $R_\sim \subseteq M \times (M/\sim)$ by $bR_\sim[c]$ if $b \sim c$. Then $R_\sim$ is a bisimulation between $\mathcal{G}_\mathcal{M}$ and $\mathcal{G}_{\mathcal{M}/\sim}$. Thus, if one maps ϵ-sets onto the equivalence class of their corresponding graphs, i.e., $\mathcal{M}(a) \mapsto [\mathcal{G}(\mathcal{M})_a]_{\text{bs}}$, then every ϵ-set is identified with each of its homomorphic images. Note that this mapping depends on the choice of the set a: If $\mathcal{M}(a) = \mathcal{M}(b)$ the corresponding graphs may not be bisimular. If we consider $[\mathcal{G}(\mathcal{M})_a]_{\text{bs}}$ as an Aczel-set in $\mathcal{A}$, it turns out that it is ϵ-isomorphic to $\mathcal{M}/\sim_{max}$. This fact is due to Sträter (1992b)):

Theorem 4. *Let* $\mathcal{M} = (M, \epsilon)$ *be an ϵ-set minimal pointed to a. Consider the ϵ-structure*

$$\mathcal{M}_{\text{afa}} := (\{[\mathcal{G}(\mathcal{M})_b]_{\text{bs}} : b \in M\}, \epsilon_{\text{afa}}).$$

1. *$\mathcal{M}_{\text{afa}}$ is an ϵ-set minimal pointed to $[\mathcal{G}(\mathcal{M})_a]_{\text{bs}}$ and ϵ-isomorphic to $\mathcal{M}/\sim_{max}$.*
2. *$\mathcal{G}(\mathcal{M}_{\text{afa}})_c$ where $c := [\mathcal{G}(\mathcal{M})_a]_{\text{bs}}$ is bisimular to $\mathcal{G}(\mathcal{M})_a$.* □

---

and thus we cannot define $\mathcal{A}_0/=_{\text{bs}}$ formally. But $\mathcal{V}$ is a model of ZFC, so $[\mathcal{G}]_{\text{bs}}$ is a class of well-founded sets (although $\mathcal{G}$ may be non well-founded as a graph), so we may consider as $[\mathcal{G}]_{\text{bs}}$ only those bisimular graphs of $\mathcal{G}$ which are of minimal rank. Then $\mathcal{A}/=_{\text{bs}}$ is a subclass of $\mathcal{V}$.

[9] By Gödel's completeness theorem a set of ε-formulas Φ is consistent iff there is a model for Φ. However, by Gödel's second incompleteness theorem, the consistency of ZFC cannot be proven within ZFC or else ZFC would be inconsistent. Thus, in ZFC, we do not know if a model of ZFC does exist. Assuming the existence of a model for ZFC (ZF, ZF$^-$ resp.) and working with this model gives so-called *relative consistency results* (see Kunen (1980) or Jech (2002)). For Gödels completeness- and incompleteness-theorems see Rautenberg (1996), for example.

The proof is straight forward: That $\mathcal{M}_{\mathrm{afa}}$ is an ϵ-set minimal pointed to $[\mathcal{G}(\mathcal{M})_a]_{\mathrm{bs}}$ follows by definition of ϵ_{afa} and the mapping $[\mathcal{G}(\mathcal{M})_b]_{\mathrm{bs}} \mapsto [b]_{\mathrm{max}}$ is an ϵ-isomorphism. For 2. one checks that $R := \{([\mathcal{G}(\mathcal{M})_b]_{\mathrm{bs}}, b) : b \in M\}$ is a bisimulation.

Hence if we consider $(\mathcal{A}, \epsilon_{\mathrm{afa}})$ as an ϵ-structure, then every ϵ-set of $\mathcal{A}$ is already factorized by the maximal extension conform equivalence. This is why the Aczel-universe is called *strongly extensional*.

We close this section with a discussion on the differences of Aczel's Theory of non well-founded sets and ϵ-style semantics. Obviously the main difference is that Aczel's Theory is a set theory alternative to ZFC, which replaces the axiom of foundation by the antifoundation axiom. In contrast to that ϵ-style semantics is not a set theory but a particular style of model theory which allows to model non well-founded relations in a set theoretical style without contradicting the axiom of foundation. The objects of interests which we use for modelling are ϵ-structures and ϵ-sets, whereas the corresponding graphs in Aczel's theory are only pictures for sets for which a set-theory is being developed.

There exist various pictures for one given set so that these pictures may be viewed as being identified within $\mathrm{ZFC}^- + \mathrm{AFA}$. Here the difference, as far as modelling is concerned, becomes visible. For example, the ϵ-sets $\mathcal{M}_{loop}$ and $\mathcal{M}_{idc}$ describe different objects in ϵ-semantics, namely a circle and an infinite descending chain of pairwise different objects (see Examles 3 and 4 above). With AFA, the corresponding apg's are bisimular and describe the same unique set. More general, given an arbitrary ϵ-set we can compare it with its homomorphic images, for example with the least or the maximal extensional quotient without identifying them, different to what would be the case in $\mathrm{ZFC}^- + \mathrm{AFA}$.

Moreover, if an ϵ-set $\mathcal{M}$ is given and if there are different $a, b \in M$ with $\mathcal{M} = \mathcal{M}(a) = \mathcal{M}(b)$ then the apg's $\mathcal{G}(\mathcal{M})_a$ and $\mathcal{G}(\mathcal{M})_b$ may fail to be bisimular and thus may be pictures for different sets while $\mathcal{M}(a)$ and $\mathcal{M}(b)$ are the same ϵ-set on which we just look from *different points of view*.

Hence, Aczel's theory has a quite different approach in using apg's for the definition of sets although these graphs are closely related to ϵ-sets and ϵ-structures.

6 $\in_T$-Logics

In this section we briefly discuss a special class of propositional logics — the so-called $\in_T$-*logics* which were motivated from the ϵ-style of semantics. Contrary to typical propositional logics the meaning $[\![\varphi]\!]$ of a formula φ in $\in_T$-logics is not given by its truth value, but as an element of a given set of propositions. Those propositions are either true or false. This can be modeled as an ϵ-structure which is generated by the classifying entities *true* and *false*; see Fig. 8.

Fig. 8.

Besides the explicit interpretation of formulas as propositions $\in_T$-logics include special predicates which state the truthness or falseness of the interpretation of a formula or which state the propositional equivalence of formulas. The proposition described by a formula (φ: true) (resp. (φ: false)) is true (resp. false) if the proposition described by φ is true (resp. false). Furthermore a formula ($\varphi \equiv \psi$) is interpreted as a true proposition if φ and ψ have the same propositional interpretation. $\in_T$-logics therefore allow the treatment of the liar paradox (the sentence *this sentence is false*) encoded in the equation

$$(x \equiv (x\text{: false})) \text{ (where } x \text{ is a variable)}.$$

Because of the fact that the liar paradox can not be stated as a true or false proposition, the liar paradox can not be expressed as a proposition in $\in_T$-logics. Thus the above equation does not have a solution in $\in_T$-logics and is treated like a conventional contradiction. The treatment of the liar paradox in $\in_T$-logics may be explained as follows: if the equation would have a solution, this solution would have to be interpreted by the liar paradox proposition, which, however, does not exist. Thus the liar paradox is not treated explicitely as a proposition in $\in_T$-logics, but is only referred to by an unsolvable equation. This allows to handle the liar paradox in $\in_T$-logics despite the totality of their truth predicates.

$\in_T$-logics are impredicative in the sense that one can quantify over the set of propositions by the quantifier $\forall$ (*for all*): a formula $\forall x.\varphi$ is a true proposition if for all propositions (represented by an assignment to the variable x) φ is being interpreted as a true proposition. This implies that quantification always includes the meaning given to $\forall x.\varphi$ as well.

Classical $\in_T$-logic is the first entity in a row of different $\in_T$-logics. It was defined by Sträter (1992a). The syntax of classical $\in_T$-logic is defined as follows:

Let $C = \{c_1, c_2, \ldots\}$ be an arbitrary set of constant symbols and $X = \{x_1, x_2, \ldots\}$ be an arbitrary set of variables. The *set $L_{\in_T}$ of formulas over C and X* is recursively defined by:

1. Every $c \in C$ and every $x \in X$ is in $L_{\in_T}$.
2. If $\varphi, \psi \in L_{\in_T}$ then so are $(\neg\varphi)$, (φ: true), (φ: false), $(\varphi \wedge \psi)$ and $(\varphi \equiv \psi)$.
3. If $\varphi \in L_{\in_T}$ and $x \in X$, then $\forall x.\varphi$ is in $L_{\in_T}$.

The semantics of classical $\in_T$-logic is defined in terms of $\in_T$-structures. It includes a set of propositions and a meaning function Γ for the interpretation of formulas.

Indeed, the definition of classical $\in_T$-logic was originally motivated from the conceptualizations of ϵ-structures (and ϵ-sets). The formal definition of $\in_T$-structures, however, is not using the concepts of ϵ-sets explicitely. Since $\in_T$-structures are just ordinary sets they do not need the extended expressiveness of ϵ-sets. More precisely, the set of propositions of an $\in_T$-structure can be seen as the union of the qdisjoint extensions of the two entities *true* and *false*, the true and false propositions. The definition of $\in_T$-structures therefore deals only with ordinary sets.

An $\in_T$-*structure* is given by $\mathcal{M} = (M, true, false, \Gamma)$ where:

1. M is a non-empty set of propositions.
2. $true \subsetneq M$ and $false \subsetneq M$ with $true \cup false = M$ and $true \cap false = \emptyset$, being the sets of true and false propositions.
3. $\Gamma : L_{\in_T} \times [X \to M] \to M$ is the meaning function *satisfying the following conditions for all formulas* φ, ψ *and all propositional assignments* β, β_1, β_2:
    a) Truth Properties:
        (1) $\Gamma((\varphi\!:\,true), \beta) \in true \Leftrightarrow \Gamma(\varphi, \beta) \in true$
        (2) $\Gamma((\varphi\!:\,false), \beta) \in true \Leftrightarrow \Gamma(\varphi, \beta) \in false$
        (3) $\Gamma((\varphi \equiv \psi), \beta) \in true \Leftrightarrow \Gamma(\varphi, \beta) = \Gamma(\psi, \beta)$
        (4) $\Gamma((\neg\varphi), \beta) \in true \Leftrightarrow \Gamma(\varphi, \beta) \in false$
        (5) $\Gamma((\varphi \wedge \psi), \beta) \in true \Leftrightarrow \Gamma(\varphi, \beta) \in true$ and $\Gamma(\psi, \beta) \in true$
        (6) $\Gamma((\forall x.\varphi), \beta) \in true \Leftrightarrow$ for all $m \in M$: $\Gamma(\varphi, \beta[x/m]) \in true$
    b) Contextual Properties:
        (1) $\Gamma(x, \beta) = \beta(x)$ for all $x \in X$ and all $\beta : X \to M$
        (2) $\Gamma(\varphi, \beta_1) = \Gamma(\varphi, \beta_2)$ if $\beta_1 \upharpoonright_{\text{Free}(\varphi)} = \beta_2 \upharpoonright_{\text{Free}(\varphi)}$
        (3) $\Gamma(\varphi[\sigma], \beta) = \Gamma(\varphi, \langle\beta\sigma\rangle)$ for all substitutions σ
        (4) $\Gamma(\varphi, \beta) = \Gamma(\psi, \beta)$ if $\varphi =_\alpha \psi$
    where Free(φ) denotes the set of free variables in φ and $\langle\beta\sigma\rangle$ denotes the propositional assignment to variables, assigning to a variable x the proposition $\Gamma(\sigma(x), \beta)$. Furthermore $\beta[x/m]$ denotes the variant of the propositional assignment β and $[\sigma]$ the substitution of formulas for σ : $X \cup C \to L_{\in_T}$. Two formulas φ, ψ are called α-congruent (written $\varphi =_\alpha \psi$) if φ and ψ only differ in the names of their bound variables.

While the truth properties in the definition of $\in_T$-structures are meant to guarantee compatibility of the meaning function to the truth functionality of the propositional operators, the contextual properties state requirements on the interpretation of formulas. The first contextual property states that the interpretation of a variable always corresponds to the assignment of the variable. The second property can be read as a coincidence statement whereby the meaning of a formula only depends on the assignment of the variables which are actually part of the formula. The third property requires a kind of compositionality by stating that the interpretation of a formula is fixed by the interpretation of its subformulas, and at the same time requires that interpretation of formulas is compatible with substitution. Formulas which only differ in the names of their bounded variables shall be interpreted equally. This is stated in the fourth property which requires that the interpretation of α-congruent formulas are equal.

Sträter (1992a) proved that there exists a sequence-style-calculus for the classical $\in_T$-logic which is sound and complete. About 10 years later Zeitz (2000) gave a simplified proof for the existence of a sound and complete Hilbert-style calculus for the classical $\in_T$-logic.

Based on Sträter's work there have been further studies in the context of $\in_T$-logics which led to several reinterpretations and generalizations. One important step in this direction has been made by Philip Zeitz (2000) who picked up Sträter's idea to study the extension of arbitrary logics by the characteristic concepts of $\in_T$-logics. To be specific, Zeitz' so-called *parameterized $\in_T$-logic* describes a methodology for the extension of parameter logics which he assumes to be given in an abstract form, by an explicit interpretation of formulas as propositions including the concepts of quantification, reference, and predicates for *true* and *false*. In the course of extension, the formulas of the parameter logic become constants in the resulting parameterized $\in_T$-logic.

Zeitz showed that the parameterized $\in_T$-logic is a generalization of Sträter's classical $\in_T$-logic in the sense that using a particular logic of constants as a parameter logic, the resulting parameterized logic exactly matches the logic of Sträter. Zeitz showed that under certain conditions sound and complete calculi of the parameter logic can uniformly be extended to become sound and complete calculi for the parameterized $\in_T$-logic.

Zeitz' main focus on developing parameterized $\in_T$-logic was to extend arbitrary logics by the propositional aspects of $\in_T$-logics. Beyond this, Zeitz's logic can also be interpreted as a methodology for the integration of logics where the resulting $\in_T$-logic forms a theory of truth and judgments over the propositions given by the integrated logics. Such a reinterpretation of Zeitz' logic can be found in Bab and Mahr (2005).

More current studies in the context of $\in_T$-logics deal with the extension of these logics by modalities taken from arbitrary modal logics whose semantics matches the terms of a possible world semantics. These studies led to the definition of a new generalization of Zeitz parameterized $\in_T$-logic, the so-called $\in_\mu$-logics. These logics describe extensions of arbitrary logics if they are given in an abstract form, by combining propositional $\in_T$ semantics with fragments of a possible world semantics. On the other hand, this generalization allows the integration of constructors coming from arbitrary modal logics. Results on $\in_\mu$-logics are presented in Bab (2007).

References

Peter Aczel. *Non–Wellfounded Sets.* CSLI Lecture Notes 14 Standford, 1988.

Sebastian Bab. Ideen zu einer modalen $\in_T$-Logik. In Sebastian Bab, Jens Gulden, Thomas Noll, and Tina Wieczorek, editors, *Models and Human Reasoning*, pages 73–88. Wissenschaft und Technik Verlag, 2005.

Sebastian Bab. $\in_\mu$-*Logik - Eine Theorie propositionaler Logiken*. Shaker Verlag, 2007. PhD-Thesis, Technische Universität Berlin, 2007.

Sebastian Bab and Bernd Mahr. $\in_T$-integration of logics. In Hans-Jörg Kreowski, Ugo Montanari, Fernando Orejas, Grzegorz Rozenberg, and Gabriele Taentzer, editors, *Formal Methods in Software and Systems Modeling*, volume 3393 of *Lecture Notes in Computer Science*, pages 204–219. Springer, 2005.

Jon Bairwise and Lawrence Moss. *Vicious Circles*. CSLI Lecture Notes 60 Stanford, 1996.

Harmut Ehrig, Bernd Mahr, Felix Cornelius, Martin Grosse-Rhode, and Philip Zeitz. *Mathematisch-Strukturelle Grundlagen der Informatik, 2. Auflage*. Springer, 2001.

Rainer Glas. *Combining Communication and Application in a λ-Like Calculus*. PhD thesis, Technische Universität Berlin, 1997.

J. Roger Hindley and Jonathan P. Seldin. *Introduction to Combinators and λ-Calculus*. Cambridge University Press, 1986.

Thomas Jech. *Set Theory - 3. millenium edition, rev. and expanded*. Springer, 2002.

Kenneth Kunen. *Set Theory - An Introduction to Independence Proofs*. North Holland, 1980.

Ralf-Detlef Kutsche. *A type oriented approach to the specification and formal semantics of a distributed, heterogeneous object system*. PhD thesis, Technische Universität Berlin, 1994.

Bernd Mahr. Applications of type theory. In J. P. Jouannaud M. C. Gaudel, editor, *Poceedings 4th Int. Conf. on Theory and Practice of Sofware Engineering (TAPSOFT'93)*, Lecture Notes in Computer Science, pages 343–355. Springer, 1993.

Bernd Mahr, Werner Sträter, and Carla Umbach. Fundamentals of a theory of types and declarations. KIT-Report 82, Technische Universität Berlin, 1990.

Moschovakis. *Notes on Set Theory*. Springer, 1994.

Ladan Pooyan. ϵ-structures as semantic models of the λ-calculus. Master's thesis, Technische Universität Berlin, 1992.

Ladan Pooyan-Weihs. *Intensional Equality for Process Calculi based on ϵ-Structures*. PhD thesis, Technische Universität Berlin, 1999.

Wolfgang Rautenberg. *Einführung in die mathematische Logik*. Vieweg, 1996.

Klaus Robering, editor. *Sorten, Typen, Typfreiheit*. Number 30 in Arbeitspapiere zur Linguistik. Technischen Universität Berlin, 1994.

Werner Sträter. $\in_T$ *Eine Logik erster Stufe mit Selbstreferenz und totalem Wahrheitsprädikat*. PhD thesis, Technische Universität Berlin, 1992a. Forschungsbericht, KIT-Report 98.

Werner Sträter. Some inofficial notes on non wellfounded sets. unpublished, August 1992b.

Eberhard Such. Bemerkungen zum Antinomienproblem. In Klaus Robering, editor, *Sorten, Typen, Typfreiheit*, pages 149–172. Technischen Universität Berlin, 1994.

Igor Tunjic. Epsilon-lambda-modelle. Master's thesis, Technische Universität Berlin, 2008.

Carla Umbach. *Termpräzisierung – Kontextuelle Steuerung der Interpretation durch Apposition und Typisierung*. PhD thesis, Technische Universität Berlin, 1996. KIT-Report 140.

Philip Zeitz. *Parametrisierte $\in_T$-Logik: Eine Theorie der Erweiterung abstrakter Logiken um die Konzepte Wahrheit, Referenz und klassische Negation*. Logos Verlag Berlin, 2000. PhD-Thesis, Technische Universität Berlin, 1999.

Assertions in AFA Set Theory

Anton Benz*

Center for General Linguistics, Berlin

benz@zas.gwz-berlin.de

Summary. In this paper we consider an application of the theory of non–well-founded sets to the modelling of epistemic updates triggered by assertions in dialogue. We concentrate mainly on what we might call *defective* dialogue, *i.e.*, dialogue in epistemic contexts where the speaker has either erroneous beliefs or is not sincere in his language use. We develop our theory within a special possible worlds framework which uses AFA set theory for representing information states of dialogue participants. We first characterise the use of assertions in *ideal* dialogue situations and then show how pragmatic principles connected to perspectives and speaker's intentions define extended uses in defective situations. On the formal level, the pragmatic considerations lead to the introduction of operators which can be used systematically to extend the domain of assertions.

1 Introduction

In a possible worlds approach the knowledge of an agent is identified with the set of all worlds where his beliefs hold true (Hintikka, 1962). In a model of communication, these worlds must contain information about the beliefs about each other. Due to the foundation axiom in set theory it is not possible to model these worlds as simple structured objects. Instead, it is necessary to use a relational model in the form of a Kripke structure (Fagin et al., 1995). Working in Aczel's (1988) set theory with anti-foundation axiom allows a direct modelling of circular and self-referential structures in terms of set membership. It is well known that this can be exploited for defining more elegant update operations which model learning effects in communication (Gerbrandy

---

* This paper is based on work done in the DFG project LA 633/5-1 on *Dialogue Semantics* located at the Humboldt–Universität Berlin. Furthermore, I have to thank the IFKI at the University of Southern Denmark and the Centre for General Linguistics, ZAS, in Berlin. The paper benefited from discussions at the Conference on Alternative Set Theories - Alternatives to Set Theory, IFKI, Kolding, 2006.

and Groeneveld, 1997). In this paper we show how this approach can be used for analysing learning effects in defective communication, *i.e.*, communication where basic epistemic constraints, as, *e.g.*, truthfulness or sincerity, are violated.

1.1 Assertions

How can we characterise the dialogue situations in which a sentence can be asserted by the speaker and interpreted by the hearer, and what do the participants learn from the fact that it has been asserted? We assume that the speaker can reasonably assert a sentence iff the epistemic update triggered by this assertion leads to a desired outcome. The hearer can interpret an assertion as long as the fact that the assertion has been made does not contradict his beliefs.

We may start with the following simple characterisation: If we identify the information which is conveyed by an assertion with the pure semantics of the uttered sentence, and if ψ represents the semantic meaning, then ψ can be asserted iff ψ is true. The interlocutors not only learn that the asserted sentence is true but also that the other participant learns that it is true, and that both know that the other one learns that it is true, *etc*, *i.e.*, it will become *mutually known* that the sentence is true. This must be represented by the update potential of assertions, and we will do this by using mutual update operations on the belief states of interlocutors.

(1) Helga calls up her son Stephan who lives in a small town in the Alps and asks him whether he wants to visit her in Munich. Stephan answers: "It is snowing in the mountains. I don't like to drive then."

Here, we consider the sentence ψ: *"It is snowing in the mountains."* Its utterance results in a situation where both speaker and hearer know that it is snowing in the mountains, and that this is mutual knowledge. But this would capture only part of the conveyed information. Helga and Stephan learn also, for example, that *Stephan knows that* ψ, and that he assumes that *Helga does not already know that* ψ. Furthermore, it should hold that *Helga believes that Stephan can know whether* ψ, and that *Stephan does not want to mislead her etc*. We call a dialogue situation where it is common knowledge that all participants have only true beliefs and that the speaker does not want to mislead the hearer an *ideal* dialogue situation. And we call an ideal situation a *basic* situation *for* ψ if all *semantic* and *pragmatic* constraints hold which are necessary for a reasonable assertion and successful interpretation.

Now let us assume that we have characterised the conditions and the update effects of assertions in ideal situations. What about non–ideal situations, *i.e.*, defective situations where beliefs are not true, or where speakers are insincere? Here the perspectives of the participants play an important role, *i.e.*, the limited information of dialogue participants and the fact that participants can exploit the limited information of other participants. In the following example of a non–basic situation both interlocutors believe that all the above mentioned pragmatic and semantic conditions hold but, in fact, the uttered sentence is not true:

(2) Helga calls up her son Stephan and asks him whether he wants to visit her in Munich. Stephan answers: "It is snowing in the mountains. I don't like to drive then." But he has not checked the weather for some time, and it is now raining and the streets are clear. ⊟

Both will update their beliefs in the same way as in the basic situation. Hence, if we have an update operation for the basic case, we can simply extend it to this wider class and get a correct description of the utterance effects. What happens if the basic conditions hold only from the perspective of the speaker?

(3) Helga calls up her son Stephan and asks him whether he wants to visit her in Munich. Stephan answers: "It is snowing in the mountains." Helga has just talked to her daughter, who lives next to Stephan, and knows that the weather has changed, and that the streets are clear. ⊟

S still feels justified to make his utterance, and the hearer may notice this. The update of the speaker remains the same as in the basic situation but the hearer should update only his representation of the speaker's beliefs. Hence we can again derive the update operation for this case from the basic one. We can think of a situation where the speaker thinks that he can mislead the hearer, and where the hearer is aware of the speaker's attempt.

(4) Helga calls up her son Stephan and asks him whether he wants to visit her in Munich. Stephan answers: "It is snowing in the mountains." Holga knows from her daughter that it does not snow at all, and she has also heard that Stephan has a new girl–friend and prefers to stay at home this weekend. ⊟

This is a situation where the utterance of *"It is snowing in the mountains"* is reasonable, and it will lead to a new update of the information states of the participants. But it is important that the hearer thinks that the speaker must believe that it is a successful lie. We can see this in contrast to **(5)**:

(5) Helga visits her son Stephan and they take a walk through the town. It is a sunny summer day, and she asks him whether he wants to take her to Munich this day. Stephan answers: "It is snowing in the mountains. I don't like to drive then." ⊟

This answer must be absolutely unintelligible for H. It can't be a false assertion, nor a lie. It can't result in a well–defined update of their belief states.

Our central assumption is: If the speaker believes that he can successfully assert some sentence, and the hearer is able to make sense of his utterance, then this is sufficient for the assertion to lead to a well–defined update of their mutual beliefs. We mean by *"being able to make sense"* of an utterance that the fact that an assertion has beed made does not *contradict* the hearer's beliefs.

The dependence on the perspectives explains why it is possible to extend the use of an assertion to new situations. On the other hand, it also leads to stronger conditions for the basic situations. Hence, there are *two* issues which we will address. The first one concerns the extension of a use to new situations, the other one the additional *pragmatic* restrictions in the basic cases. In addition to perspectives we consider speaker's goals. We need them in order to introduce a *sincerity* condition for the base case, for example, to explain the difference between (**2**) and (**3**).[2]

1.2 Non–well–founded sets

In Section 2 we introduce our general framework. We represent dialogue situations and the beliefs of dialogue participants within a special *possible worlds* framework (Sec. 2). We will build up upon an the approach developed by *Gerbrandy* and *Groeneveld* (1997). Possible worlds w, or *possibilities*, are triples $\langle s_w, w(S), w(H) \rangle$. The first component represents the situation talked about, the second and third the belief sets of speaker and hearer. The sets $w(S)$ and $w(H)$ are again sets of possibilities. As they are proper parts of w, standard set theory would not allow that w is itself an element of $w(S)$ or $w(H)$. But clearly, if the belief sets collect all possibilities that an agent believes to be possible, then a belief set $w(X)$ can only represent a *true* belief iff $w \in w(X)$. In standard possible worlds frameworks (Hintikka, 1962; Fagin et al., 1995), this problem is solved by using relational models, *i.e.*, an agent believes in w that a world v is possible iff w stands in a certain *accessibility relation R* to v. A disadvantage of this approach is that the belief sets, here the set of all worlds accessible to an agent, do no longer enter into the identity conditions of worlds. Using AFA set theory (Aczel, 1988; Barwise and Moss, 1996) we can solve both problems, *i.e.*, possibilities w can have belief sets which contain w, and possibilities are identical if, and only if, their situations talked about and their belief sets are identical. In AFA set theory, knowledge updates can be reduced to simple set theoretic manipulations of possibilities, which leads to elegant and perspicuous mathematical representations (Gerbrandy and Groeneveld, 1997).

1.3 Assertions in AFA set theory

In Section 3 we present our basic considerations concerning the role of perspectives (Sec. 3.1) and speaker's goals (Sec. 3.2) in dialogue. There we introduce

---

[2] In (**3**) Helga can interpret Stephan's utterance as a lie only because she knows that he does not want to visit her, hence the result of a successful lie would fit his goals. In Example (**2**) she can infer that his utterance is due to an error because she assumes that he is sincere.

operators linked to the perspectives of speaker and hearer which allow us to *derive* extended classes of situations where the speaker believes to be justified in making an assertion and where the hearer can make sense of an assertion. If M is a class where the use of an assertion is well–defined, then we can apply the operators for speaker and hearer to M and get the classes where from the perspective of speaker or hearer the assertion is justified. One of the major problems is to characterise the belief–structures of the dialogue participants in situations which are possible candidates for an extended use, *i.e.*, to characterise the class of dialogue situations where it makes sense to ask whether an extended use of an assertion is possible. In this regard, we introduce a class of possibilities with *internal hierarchical* structure, $\mathcal{H}(\mathcal{T})$, in Section 3.3. The motivation for this structure is closely related to the role of perspectives in deriving extended uses. This defines the class of intended applications of our theory.

In Sections 4 and 5 we show in detail how to derive the basic cases and extended uses. We start with the mutual update operations which are defined by the pure semantics of an assertion in ideal dialogue situations. We consider two groups of pragmatic principles: a) principles linked to perspectives, and b) rationality principles linked to speaker's goals. We will divide the main body of our investigation (Sec. 4/5) into two parts. In the first part we consider only principles linked to perspectives and their interaction with the pure semantics of an assertion (Sec. 4). This part again divides into a sub–part (Sec. 4.1) where we define the pragmatically basic situations where a sentence can be asserted, and a second sub–part (Sec. 4.2) where we show how perspectives give rise to extended uses. The principles linked to perspectives provide for the justification of these extended uses and allow us to define the update operations for a derived use of an assertion in terms of the update operation which represents the underlying use. In the second part we discuss effects of rationality principles connected to speaker's goals (Sec. 5). This section again divides into a sub–part (Sec. 5.1) where we consider the base case problem, and a sub–part (Sec. 5.2) where we consider derived uses of assertions. The underlying circularity of structures and conditions makes it clear that this enterprise is by no means trivial. In Section 6 we show how the examples are handled using our theory.

2 Possibilities in AFA set theory

The possibility approach is essentially a possible worlds approach, *i.e.*, it identifies the beliefs of an individual with the set of all worlds which are *possible* according to those beliefs. We denote the set of participants by $\mathrm{DP} = \{S, H\}$, where S denotes the *speaker*, H the *hearer*. A *possibility* consists of a model for the *outer* situation and *information states* for each participant, where those states are again sets of possibilities. The *outer* situation describes the *non–modal* part of the dialogue context. In case of assertions, for example, this outer situation will be identified with the situation *talked about*. The possibility approach was first developed by J. *Gerbrandy* and W. *Groeneveld* in (1997). It is based on an extension of classical set theory, the theory of *Non–Well–Founded Sets* developed by P. *Aczel* (1988).[3] The original problem motivating the development of the possibility approach was to define suitable *update operations* for dialogue. Here the approach proved to be especially useful.[4] Before we introduce possibilities, we first state some of the fundamental theorems of AFA set theory.

AFA set theory

For Cantor or Frege, a *set* was an unrestricted collection of objects that satisfy a certain predicate. Understood thus, a set can be seen as the semantic extension associated to a predicate. In an effort to avoid the paradoxes generated by unrestricted comprehension, the cumulative conception of the set universe emerged.[5] Intuitively, we can think of the set universe as generated from base sets by iterated application of a power set operation. The set universe is organised in levels where each higher level comprises all subsets of the lower levels. In AFA set theory the emphasis shifts away from the cumulative conception. Here, the aspect that is relevant to us is the conception of a set as a structured object where the identity of objects follows from the identity of its constituents, *i.e.*, elements.

AFA set theory has no axiom of foundation. The following so-called *Solution Lemma* can either be proved as a theorem in AFA set theory (Aczel, 1988) or be introduced as an axiom (Barwise and Moss, 1996). In the latter case, it is the axiom which replaces the axiom of foundation:

Theorem 2.1 (Solution Lemma) *Let $\mathbf{V}[X]$ be the universe of sets over a class X of urelements. A system of equations over X is a function $e: X \longrightarrow \mathbf{V}[X] \setminus X$. Then, every system of equations has a unique solution; i.e., there exists a unique function $s: \mathbf{V}[X] \longrightarrow \mathbf{V}$ such that*

$$s(m) = \{s(n) \mid n \in m\} \quad \text{for sets } m,$$
$$s(x) = s(e(x)) \quad \text{for urelements } x \in X.$$

To see the significance of this lemma, let us consider an arbitrary set m. We can associate to m the set of equations $n = \{k \mid k \in n\}$, where n is in the transitive hull of m. In a second step we can replace each set n by a distinct urelement u_n, which leads to the system of equations $u_n = \{u_k \mid k \in$

---

[3] For more information about (AFA) set theory we can refer to (Barwise and Moss, 1996), and for the possibility approach to the thesis of Gerbrandy (1998).
[4] There is a wider range of literature concerning the proper definition of updates in communication: (Jaspars, 1994), (Groeneveld, 1995), (Gerbrandy and Groeneveld, 1997), (Zeevat, 1997), (Gerbrandy, 1998), (Balta et al., 1999), (Baltag, 1999), (Baltag et al., to appear) and more.
[5] For the history of this development see (Fraenkel and Bar-Hillel, 1958).

$n\} =: e(u_n)$. A solution to this system of equations is a function s that assigns to each urelement u_n a set $s(u_n)$ such that $s(u_n) = \{s(u_k) \mid k \in n\} = \{s(e(u_k)) \mid k \in n\}$. We can think of the system of equations as a description of the internal structure of a set. In standard set theory every set has such a structural description, and if a system of equations has a solution, then this solution is unique. AFA set theory generalises this property by assuming that *every* system of equations has a unique solution. Two sets are identical if they can be described by the same system of equations, i. e, if they have the same structural description. The following identity condition follows from the solution lemma:

Theorem 2.2 *A relation $R \subseteq \mathcal{W} \times \mathcal{W}$ is a bisimulation, iff vRw implies that $s_v = s_w$ and for all $X \in \mathrm{DP}$*

$$\forall v' \in v(X) \, \exists w' \in w(X) \, v'Rw' \text{ and } \forall w' \in w(X) \, \exists v' \in v(X) \, v'Rw'.$$

If R is a bisimulation, then vRw implies $v = w$. □

Possibilities

In order to define certain subclasses of non–well–founded sets we need the following special fixed–point lemma:

Lemma 2.3 *For each class M there exists a unique maximal class C which satisfies:*

$$C = \{\langle m, x, y\rangle \mid m \in M \, \& \, x, y \subseteq C\} \quad (*)$$ □

C can be constructed as a fixed-point of the operator $\Gamma X := \{\langle m, x, y \rangle \subseteq X \mid m \in M \, \& \, x, y \subseteq X\}$. Let $C^0 := \mathbf{V}$, $C^{n+1} := \Gamma \, C^n$, $C := \bigcap_n C^n$, then C is a fixed–point of Γ and the largest class that satisfies $(*)$.[6]

Let $\mathcal{M}$ be a class of models for the possible outer situations. We define *possibilities* and *information states* in the following way:

- A *possibility* w is a triple $\langle s_w, w(S), w(H)\rangle$ in which $s_w \in \mathcal{M}$ and $w(S)$ and $w(H)$ are information states.
- An *information state* σ is a set of possibilities.

s_w describes an outer situation, $w(S)$ and $w(H)$ the set of worlds S and H believe to be possible. We denote the class of all possibilities with $\mathcal{W}$. The theory of non–well–founded sets allows for sets containing themselves, hence it is possible that there exist possibilities w with $w \in w(X)$, $X \in \mathrm{DP}$.

Let $\mathcal{L}$ be a language of predicate logic for the class $\mathcal{M}$. We assume that $\mathcal{L}$ contains all the predicates the dialogue participants can use to talk about an outer situation. Let $w = \langle s_w, w(S), w(H)\rangle$ be a possibility. We define truth conditions for $\varphi \in \mathcal{L}$:

---

[6] For the set–theoretic machinery behind fixed–point constructions we refer to (Aczel, 1988; Barwise and Moss, 1996).

$$w \models \varphi \text{ iff } s_w \models \varphi, \ \varphi \text{ a sentence in } \mathcal{L}.$$

For a dialogue participant X a possibility w is *epistemically possible* in v iff $w \in v(X)$. X *believes* that φ in w iff φ holds in all his epistemic alternatives in w. We write $w \models \Box_X \varphi$ iff $\forall v \in w(X)\, v \models \varphi$, and $w \models \Diamond_X \varphi$ iff $\exists v \in w(X)\, v \models \varphi$. If σ is an information state, then we define $\sigma \models \varphi$ iff $\forall w \in \sigma\, w \models \varphi$.

Until now, we did not restrict the properties of possibilities. A subclass $M \subseteq \mathcal{W}$ is called *transitive*, iff $\forall w \in M\, \forall X \in DP\, w(X) \subseteq M$. Let $\mathcal{I} \subseteq \mathcal{W}$ be the largest transitive subclass with

$$\forall w \in \mathcal{I}\, \forall X \in DP\, \forall v \in w(X): \ w(X) = v(X).$$

This property is called *introspectivity*. It means: (1) If a dialogue participant believes φ, then he knows that he believes it; (2) if he does not believe that φ, then he knows that he does not believe φ; and (3) it means that (1) and (2) are common knowledge. Let $\mathcal{T} \subseteq \mathcal{I}$ be the largest transitive subclass with

$$\forall w \in \mathcal{T}\, \forall X \in \mathrm{DP}\, w \in w(X).$$

If $w \in \mathcal{T}$, then w is for both participants an element of their sets of epistemic alternatives. Hence, if a participant believes that φ, then φ must in fact hold. Therefore, $\mathcal{T}$ denotes the class of possibilities in which (1) the dialogue participants can only have *true beliefs*, i.e., *knowledge*, and (2) in which this fact is common knowledge. The *Anti–Foundation–Axiom* (AFA) of the underlying set theory guaranties that $\mathcal{T}$ is not empty.

We are only interested in non–contradicting information states of participants. This means that the set containing all their epistemic possibilities should contain at least one element. Let $\dot{\mathcal{I}}$ denote the largest transitive subclass of $\mathcal{I}$ with $w \in \dot{\mathcal{I}} \Rightarrow w(S) \neq \emptyset \neq w(H)$.

If $D(a) \subseteq \mathcal{W}$ and $a: D(a) \longrightarrow \mathcal{W}$, then we call a a *normal mutual update* if

(**NMU**) $\quad a(w) = \langle s_w, a(w(S)), a(w(H)) \rangle$ for $w \in D(a)$,
$\qquad\qquad a(\sigma) = \{a(w) \mid w \in \sigma \cap D(a)\}$ for information states σ.

It is clear that a normal mutual update is uniquely determined by its domain $D(a)$. If $[\varphi] := \{w \in \mathcal{T} \mid w \models \varphi\}$, then the normal mutual update determined by $[\varphi]$ describes the effect of '*mutually learning that φ*' in $\mathcal{T}$. We can think of this mutual learning as a step by step process of eliminating possibilities. First both participants eliminate all possibilities in which φ is not true from their information states, and then they take the remaining possibilities and again eliminate all worlds from both information states in which φ is not true. This they repeat again and again. Such definitions of update operations for possibilities have been introduced by (Gerbrandy and Groeneveld, 1997). From the set theoretic point of view, (**NMU**) defines a system of equations, i.e., a structural description of the possibility that results from updating.

3 The Basic Considerations

In this section we describe the basic ideas of our approach and provide a definition of the class of possibilities which represent the intended applications. We have already explained that we make a distinction between assertions in ideal and non–ideal dialogue situations. We represent the class of ideal situations by $\mathcal{T}$, i.e., we assume that all participants have only true beliefs, and that this is common knowledge. The *basic* assertions of a sentence ψ are all uses of ψ in $\mathcal{T}$ where all pragmatic and semantic constraints hold. We will denote this class by $\mathcal{B} = \nabla[\psi] \subseteq \mathcal{T}$. We can represent the effect of mutually learning ψ by a normal mutual update $a : [\psi] \longrightarrow \dot{\mathcal{I}}$, where $[\psi] := \{w \in \mathcal{T} \mid w \models \psi\}$. But if ψ is really asserted, then both participants not only learn that ψ holds, but also that all pragmatic conditions hold, i.e., they learn that they are in a situation in $\mathcal{B}$. This means that we can identify an assertion with the normal mutual update $a : \mathcal{B} \longrightarrow \dot{\mathcal{I}}$. We then want to extend this basic use to non–ideal situations. We will see that it is not possible to identify these non–ideal cases with the complement $\dot{\mathcal{I}} \setminus \mathcal{T}$. This forces us to describe the intended applications more closely. Bearing this in mind, we introduce in Section 3.3 a class $\mathcal{H}(\mathcal{T})$. In the same way as $\mathcal{T}$ represented the class of situations where it was reasonable to ask whether a basic use is possible, $\mathcal{H}(\mathcal{T})$ will represent the class where it is reasonable to ask whether a derived extended use is possible. We extend $\mathcal{B}$ to a class $\Delta(\mathcal{B}) \subseteq \mathcal{H}(\mathcal{T})$ by iterated applications of certain operators which are linked to perspectives. They reflect the way perspectives restrict and allow use and interpretation of assertions. Later (Sec. 4.1) we will see that we can also derive $\mathcal{B} = \nabla[\psi]$ by an iterated application of a related combination of operators to $[\psi]$. We first (Sec. 3.1) motivate these operators and then characterise the non–ideal dialogue situations in which we intend to apply our theory.

3.1 Considerations concerning Perspectives

The Operators

We reconsider the assertion *It is snowing in the mountains* ψ in Example (1). If we define the felicity conditions by pure semantics, then asserting ψ would be felicitous just in case the sentence ψ is true. We represent the situation of Example (1) by a possibility w with[7]:

$$w = \langle \psi, \{w\}, \{w, v\} \rangle$$
$$v = \langle \neg \psi, \{v\}, \{w, v\} \rangle.$$

i.e., w is such that ψ holds, the speaker knows the actual situation, and the hearer can not distinguish between w and v, where the only difference between

---

[7] As we are only interested in the truth or falsity of ψ we denote the outer situation just by ψ or $\neg \psi$.

w and v is that $\neg\psi$ holds in v. Hence the semantic condition for asserting ψ holds. But, of course, the speaker S can rely for what he says only on what he believes to be true. We assume throughout this paper that the speaker is only justified to assert a sentence if he is *convinced* that he can assert it. For the basic case this means that $w(S)$ should be a subset of all situations where ψ holds. This is, of course, in many real life situations too strong a requirement, as agents may well perform an action if they think, *e.g.*, that the possible negative results are negligible, or that it is highly unlikely that it has no success. If $M \subseteq \dot{\mathcal{I}}$ is a class which represents some property of possibilities, *e.g.*, a class where some sentence can be asserted successfully, then we explicate the fact that this property obtains *under the perspective* of a dialogue participant X in world w as $w(X) \subseteq M$.

Now we look at the hearer H. We see that one of his epistemic alternatives is that ψ does not hold. It is also essential that the addressee does not already know that the uttered sentence is true. Hence, if M characterises the pragmatic conditions for a reasonable use of an assertion, then the requirement that he must believe that all his epistemic alternatives are situations in M is too strong. We assume that he *can make sense* out of an assertion if there is at least one epistemic alternative in his set of possibilities where the uttered sentence is reasonable. *Make sense* means here that if the hearer *learns* that a sentence ψ is asserted, then this fact does not contradict with what he believes about the world. The actual situation w *may* have the property represented by a class M *under the perspective of* a participant X iff $w(X) \cap M \neq \emptyset$.

We define two operators on subclasses of possibilities for each dialogue participant X. They are closely related to the modal operators $\Box_X$ and $\Diamond_X$, so we denote them by the same symbols:

$$\Box_X M := \{w \in \dot{\mathcal{I}} \mid w(X) \subseteq M\}$$
$$\Diamond_X M := \{w \in \dot{\mathcal{I}} \mid w(X) \cap M \neq \emptyset\}$$

With these operators at hand we can reformulate our observations as: S is convinced that the actual world w belongs to M iff $w \in \Box_S M$. Learning that M does not contradict what H believes iff $w \in \Diamond_H M$. Notice that these operators generate proper classes, even if M is a set. Our aim is to derive the class of extended uses of an assertion by an iterated application of certain combinations of these operators to the base $\mathcal{B}$.

Combinations of Operators for Derived Uses

Let us now look at Example **(2)**. We represent it by

$$u = \langle \neg\psi, \{w\}, \{w, v\}\rangle,$$

where w and v are the situations from Example **(1)**. We see that $u(S) = w(S)$ and $u(H) = w(H)$. If the assertion leads to an update in w, then it should also do so in u. If we assume that we have proven that w belongs to the basic cases for a successful use, and if we assume that we have characterised these basic uses by a class $\mathcal{B}$, then we see also that $u, w \in \Box_S \mathcal{B} \cap \Diamond_H \mathcal{B}$.

In general we assume that an assertion leads to a successful update if the speaker is convinced that he can use the assertion, and if the fact that it is asserted does not lead to a contradiction with the beliefs of the hearer. Hence, if we have proven that an assertion leads to an update for a class of situations M, then it also leads to an update of the speaker's information state for situations in $\Box_S M$, and to an update of the hearer's information state for situations in $\Diamond_H M$.

If the speaker thinks that the hearer can make sense of an assertion, he can exploit this fact and mislead him. This happens in case of lying. We can see here how the limited perspective of one dialogue participant can give rise to an extension of a dialogue act. If M is a class of possibilities where an assertion leads to an update, then this act can be extended to the class where S is convinced that they are in a situation in M or where H can at least make sense of the assertion. This is the class $\Box_S(M \cup \Diamond_H M)$. If $M \subseteq \Diamond_H M$, the definition of the extension can be simplified to $\Box_S \Diamond_H M$. This will always be the case.

Now we can again turn to the perspective of H. Assume that the speaker S is convinced of the truth of ψ while H knows it to be false. This was the case in Example (**3**) which we can represent by:

$$s = \langle \neg \psi, \{w\}, \{s\} \rangle,$$

where w is the situation from Example (**1**). We see that $s \in \Diamond_H \Box_S \mathcal{B}$, where $\mathcal{B}$ denotes the class of basic situations for the assertion ψ (*It is snowing in the mountains*). We assume H can make sense of the utterance if he thinks it is possible that S might believe that he can assert ψ. Let M be again a class of possibilities where we know that the assertion leads to an update. Then the assertion also leads to an update of the hearer's information state if H thinks that it is possible that S is convinced that the actual situation belongs to M. I.e., there exists an update of the hearer's information state for the class $\Diamond_H \Box_S M$.

We find in this way four operations which give us new classes in which sentence can be asserted due to the perspective of *one* participant. Let M be given. Then we classify the perspectivally derived classes as shown in Figure 1.

	direct	indirect
speaker	$\Box_S M$	$\Box_S(M \cup \Diamond_H M)$
hearer	$\Diamond_H M$	$\Diamond_H \Box_S M$

Fig. 1. The four basic perspectival operations.

We can simplify $\Box_S(M \cup \Diamond_H M)$ to $\Box_S \Diamond_H M$ in the case of an indirect operation for the speaker if $M \subseteq \Diamond_H M$. We will see later that this is the case in all intended applications.

It follows by *introspectivity* that we can not get new possible extensions if we apply the direct operation twice for the same participant. I.e., for operators $P, Q \in \{\Box, \Diamond\}$, $M \subseteq \dot{\mathcal{I}}$ and $X \in \mathrm{DP}$ we find that $P_X Q_X M = Q_X M$.

To arrive at a real extension it is necessary that S is convinced that he can perform the act, and H must be able to make sense of this. It is not sufficient that only one participant thinks that the act can be performed. Therefore, we have to build intersections of the derived classes. We get the four groups of derived cases as shown in Figure 2.

	direct H	indirect H
direct S	$\Box_S M \cap \Diamond_H M$	$\Box_S M \cap \Diamond_H \Box_S M$
indirect S	$\Box_S \Diamond_H M \cap \Diamond_H M$	$\Box_S \Diamond_H M \cap \Diamond_H \Box_S M$

Fig. 2. The four derived classes.

3.2 Considerations concerning Speaker's Goals

So far we have not taken speaker's intentions into account. In Example (**4**) Helga will interpret Stephan's utterance as a lie because she knows that it fits his intentions. This contrasts with Example (**3**) where she will interpret the same utterance as a mistake because she has no reason to assume that he wants to mislead her. Also in a basic situation like in Example (**1**) the hearer not only learns that some fact ψ holds and that the speaker believes in ψ, but also that he wants to inform her about this fact. The following example is identical with Example (**1**) except for the fact that it is common knowledge that the speaker wants to mislead the hearer in one possible context:

(**6**) Helga calls up her son Stephan. The last time she invited him, he pretended that he could not come because of the bad weather, and she knows that he does not like to come this time either. They both know that Stephan would like to use the same excuse this time. ⌑

If we don't consider speaker's goals, then this situation receives the same representation as Example (**1**). But, of course, if Stephan says that *it is snowing in the mountains*, then this should lead to a different update this time. He should not be able to convince Helga. This means that our theory should predict that it is not rational to make the assertion in this context. To this end we introduce speaker's intentions and rationality constraints connected to goals. We represent speaker's goals by a function which gives us the desired states of the world for all his epistemic alternatives, i.e., we redefine possibilities as triples of the form $\langle s_w, \langle w(S), G_S^w \rangle, \langle w(H) \rangle \rangle$ where G_S is a function with domain $w(S)$ and values $G_S(v) \subseteq \dot{\mathcal{I}}$.[8]

Our model does not allow to rank possible outcomes, so we can't expect it to provide a real criterion for rational *choice*. Our goal–functions divide the class of all possibilities into the class of possibilities where the goal is achieved, and the class where it is not achieved. We formulate criteria which tell us whether it is reasonable to choose an action as a *means* to reach the goal.

We formulate two elementary constraints which tell us whether it is rational to assert a sentence relative to some goal G. Assume that there are given a situation w, a mutual update a and a goal $G(w)$. Let w be in the domain of a. Then we postulate the following rationality constraints:

(**R$_1$**) $w \notin G(w)$
(**R$_2$**) $a(w) \in G(w)$

The first axiom claims that we should only perform an action if our goal is not yet achieved. If it is, nothing should be done. The second axiom states that we should only choose actions which allow us to reach our goal, *i.e.*, the result $a(w)$ of the performance of a in w should be in $G(w)$. We can also see that (**R$_1$**) and (**R$_2$**) imply that $a(w) \neq w$. We can incorporate these constraints in our operators, but this will have to wait until Section 5. If (1) we know that there is a class M where we have already established that a felicitous use of an assertion is possible, and if (2) w is an element of one of the classes which can be derived by a combination of perspectival operators for speaker and hearer, and if (3) the update with the information connected to the fact that ψ has been asserted, *i.e.*, with $\Delta(\mathcal{B})$, satisfies the rationality conditions, then this *proves* that a felicitous use of the assertion is also possible in w. (3) makes the criterion circular.

3.3 Characterising the Intended Applications

We start with the felicity conditions and update effects of assertions as they are defined by pure semantics, *i.e.*, we assume that a sentence with semantic content ψ can be asserted exactly iff the actual situation belongs to $[\psi] := \{w \in \mathcal{T} \mid w \models \psi\}$. Its effect is described by the normal mutual update with this semantic content, *i.e.*, by $a : [\psi] \longrightarrow \dot{\mathcal{I}}$. The pragmatically *basic* cases build the subclass $\mathcal{B}$ of $[\psi]$ in which the additional pragmatic constraints linked to perspectives and intentions hold. We assume here (1) that the speaker is sincere and must know that he is successful, and (2) that the hearer must be able to make sense of the utterance, *i.e.*, updating with the information which is connected to the fact that ψ was uttered does not lead to a contradiction. The implicit circularity of these constraints makes the task of finding $\mathcal{B}$ a non–trivial one.

---

[8] The precise definition will follow in Section 5.

We extend $\mathcal{B}$ to non–ideal dialogue situations by applying a combination of the perspectival operators. Let us assume now that we have applied the extending operation α times and got $\Delta^\alpha(\mathcal{B})$. $\Delta^\alpha(\mathcal{B})$ should also represent the information which is connected to the fact that ψ was uttered, and we can identify the update potential with the normal mutual update $a : \Delta^\alpha(\mathcal{B}) \longrightarrow \dot{\mathcal{I}}$ defined by $\Delta^\alpha(\mathcal{B})$. What we want is to define $\Delta^{\alpha+1}(\mathcal{B})$ as $\Delta^\alpha(\mathcal{B}) \cup D(\Delta^\alpha(\mathcal{B}))$ where D denotes the operation which derives extended uses. Then we can identify the new update effect of asserting ψ with the normal mutual update defined by $\Delta^{\alpha+1}(\mathcal{B})$. In the end we want to collect all $\Delta^\alpha(\mathcal{B})$ in a class $\Delta(\mathcal{B})$. This then proves that the pragmatic principles connected to perspectives and the rationality constraints allow extended uses of the assertion of ψ exactly iff the utterance situation belongs to $\Delta(\mathcal{B})$. But this does not work out so smoothly. For instance, we have said that $\mathcal{B}$ should characterise the pragmatically basic uses, i.e., a sentence ψ can be asserted in an ideal dialogue situation $w \in \mathcal{T}$ if, and only if $w \in \mathcal{B}$. But if we now apply our extending operators, this may add new elements from $\mathcal{T}$, and this first contradicts the claim that we can assert ψ in $\mathcal{T}$ only if it is an element of $\mathcal{B}$, and second the update of a situation in $\mathcal{B}$ with the larger set $\Delta^{\alpha+1}(\mathcal{B})$ may now lead to a different result. This leads to problems in connection with the rationality constraints, Example (7) below. We solve these problems by restricting our applications to a class $\mathcal{H}(\mathcal{T})$ of candidates with an internal hierarchical structure. There is a *hard* reason for introducing $\mathcal{H}(\mathcal{T})$ which is related to speaker's intentions. If we allow for possibilities with arbitrary circular structure, then we find situations where our criterion for extended uses leads into an irresolvable circle:

(7) Let w be the situation from Example (1), i.e., $w = \langle \psi, \{w\}, \{w, v\} \rangle$ and $v = \langle \neg\psi, \{v\}, \{w, v\} \rangle$. Then we consider
$$t = \langle \neg\psi, \{t\}, \{t, u\} \rangle,$$
$$u = \langle \psi, \{w\}, \{t, u\} \rangle.$$
□

We further assume that in t and w the speaker wants to convince the hearer that ψ holds. Furthermore, we assume that in w it is common knowledge that the speaker is sincere. Hence we assume $w \in \mathcal{B}$. Now we are faced with the following problem: Neither t nor u are elements of $\mathcal{T}$, hence, in the initial situation no use of an assertion is defined. Then we try to extend $\mathcal{B}$ step by step. First we see that $u \in \Box_S \mathcal{B} \cap \Diamond_H \Box_S \mathcal{B}$. It follows that a derived use is possible in u. As t and u do not belong to $\mathcal{B}$ it follows that t is not contained in the first extension $\Delta^1(\mathcal{B})$. Hence the related update with $\Delta^1(\mathcal{B})$ should eliminate t from H's information state in u. But this means that the update results in a state where H is convinced that ψ holds. Now we apply our operators once more, and we see that $t \in \Box_S \Diamond_H \Delta^1(\mathcal{B}) \cap \Diamond_H \Delta^1(\mathcal{B})$. As S can also see that an update with $\Delta^1(\mathcal{B})$ leads to a desired situation, it follows that $t \in \Delta^2(\mathcal{B})$. But now the mutual update with $\Delta^2(\mathcal{B})$ does not remove t from H's information state, hence, she will no more be convinced that ψ holds. This means that we should remove t again if we define $\Delta^3(\mathcal{B})$ because a rationality condition is violated. But then the update with $\Delta^3(\mathcal{B})$ eliminates t in $u(H)$ and we are in the same situation we have been in after the first derivation.

We are going to characterize the intended applications by a class $\mathcal{H}(\mathcal{T})$. The elements of $\mathcal{H}(\mathcal{T})$ will have an *internal hierarchical structure* which allows us to measure their complexity by ordinal numbers. We then construct $\Delta^{\alpha+1}(\mathcal{B})$ in such a way that our operators add only elements with complexity $\alpha+1$. Then our claim will be that a sentence ψ can be asserted in a situation $w \in \mathcal{H}^{\alpha+1}(\mathcal{T})$, i.e., in situations in $\mathcal{H}(\mathcal{T})$ with complexity at most $\alpha+1$, if and only if $w \in \Delta^{\alpha+1}(\mathcal{B})$. $\mathcal{H}(\mathcal{T}) \setminus \mathcal{T}$ is the class of candidates for a derived extended use in the same way as $\mathcal{T}$ was for the basic uses.

For the direct derivation, our intuition has been that the participant believes to be in a situation where some sentence can be asserted successfully. For the indirect case, he has to be convinced that the other one is or might be in such a situation. In both cases, we think that the possibilities in the exploited information states have to be more simple than the newly derived possibilities. In the ideal case of Example (1) we see that the hearer not only believes that they may be in a basic situation $w \in \mathcal{B}$ for ψ but also that she is *convinced* to be in an ideal situation, i.e., the actual situation belongs to $\Diamond_H \mathcal{B} \cap \Box_H \mathcal{T}$. In Example (4) the speaker believes that he can mislead the hearer. But this is only the case because he is convinced that the hearer is convinced to be in an ideal situation, i.e., the actual situation belongs to $\Box_S \Diamond_H \mathcal{B} \cap \Box_S \Box_H \mathcal{T}$. In Example (3) the hearer interprets the assertion as a mistake because she is convinced that he acts according to the conditions of ideal dialogue, i.e., the actual situation is in $\Diamond_H \Box_S \mathcal{B} \cap \Box_H \Box_S \mathcal{T}$. In the case of Example (7) we see that $u \in \Diamond_H \mathcal{B}$ but $u \notin \Box_H \mathcal{T}$. These observations lead us to the following generalisation: If we have established that a sentence ψ can be asserted in situation $w \subset G$ if, and only if $w \models M$, then we find the *candidates* for an extended use by looking at the table in Figure 3: This

	direct candidate		indirect candidate	
speaker	$\Box_S M$	$\Box_S G$	$\Box_S(M \cup \Diamond_H M)$	$\Box_S(G \cup \Box_H G)$
hearer	$\Diamond_H M$	$\Diamond_H G$	$\Diamond_H \Box_S M$	$\Diamond_H(G \cup \Box_S G)$

Fig. 3. The four derived uses and their candidate classes.

means, if an extended use is possible in a situation v, then v has to be, e.g., an element of $(\Box_S M \cap \Box_S G) \cap (\Diamond_H \Box_S M \cap \Diamond_H \Box_S G)$. The operators in the second and fourth column provide us with the candidates where speaker and hearer can search for extended uses. We can construct the class $\mathcal{H}(\mathcal{T})$ by an iterated application of these operators. We will prove in Lemma 4.3 and 5.2 that, in fact, the problems of Example (7) do not occur if we restrict our applications to this class.

We now look for a description which can be better used in the subsequent proofs. For this reason, we provide for an *axiomatic* characterisation of $\mathcal{H}(\mathcal{T})$.

In Section A, Lemma A.7, we will prove that the recursive construction and the axiomatic characterisation define the same structure $\mathcal{H}(\mathcal{T})$. The construction by an iterated application of the above operators suggests that we can characterise $\mathcal{H}(\mathcal{T})$ in terms of *paths*. And, in fact, we can do this: $\mathcal{H}(\mathcal{T})$ is the largest transitive subclass of $\dot{\mathcal{I}}$ such that for all situations $w \in \mathcal{H}(\mathcal{T}) \setminus \mathcal{T}$:

1. There is at most one participant who believes the real situation to be possible.
2. There are no long circular paths going from one participant to the other and coming back to the original situation.
3. If there is a path starting at the real situation which goes from one participant to the other, then this path should ultimately reach the ideal situations in $\mathcal{T}$.

i.e., we don't allow for structures where we have $(\mathbf{H_{1'}})$ $w \in w(S) \cap w(H)$, or $(\mathbf{H_{2'}})$ sequences $v_1 \in v_0(X_0) \& v_2 \in v_1(X_1), \ldots, v_0 \in v_n(X_n)$ where $v_0 \in \mathcal{H}(\mathcal{T}) \setminus \mathcal{T}$, for all i it holds that $X_i \neq X_{i+1}$ and $n > 0$, or $(\mathbf{H_{3'}})$ sequences $v_1 \in v_0(X_0) \& v_2 \in v_1(X_1), \ldots$ where for all i holds that $X_i \neq X_{i+1}$ and where no $v_i \in \mathcal{T}$. These quite intuitive conditions may help to understand our final characterisation of $\mathcal{H}(\mathcal{T})$. We will see in Section 2 that the following axioms $(\mathbf{H_1})$–$(\mathbf{H_3})$ capture the content of $(\mathbf{H_{1'}})$–$(\mathbf{H_{3'}})$. We provide for a more fine–grained structure because it allows us do define extended uses by recursion over the *complexity* of situations.

Let $T(w)$ be the smallest transitive superset of $\{w\}$. We call $T(w)$ the *transitive hull* of the possibility w. In the case of Example **(1)** $T(w)$ is $\{w, v\}$, for Example **(2)** $T(u) = \{u, w, v\}$, and for Example **(3)** $T(s) = \{s, w, v\}$. We can see that $T(w) \subseteq T(u), T(s)$ but $T(u), T(s) \not\subseteq T(w)$. In general, we find, e.g., for $w \in \square_S \mathcal{T} \cap \square_H \mathcal{T}$ with $w \notin \mathcal{T}$ that for all $v \in w(S) \cup w(H)$: $T(w) \not\subseteq T(v)$. In the case of Example **(7)** we see that $T(t) = T(u)$.

In a first step we restrict the intended applications to cases where the subset relation between transitive hulls defines a well–founded partial order on dialogue situations:

- $[w] := \{v \in \dot{\mathcal{I}} \mid T(w) = T(v)\}$,
- $[v] \leq [w]$ iff $T(v) \subseteq T(w)$.

Let $\mathcal{F}$ be the class of all possibilities $w \in \dot{\mathcal{I}}$ where $\leq$ is a well–founded partial order on $\{[v] \mid v \in T(w)\}$. We can see that Example **(7)** is still an element of $\mathcal{F}$. For $\mathcal{F}$ we can define an *order type* for the possibilities. This order type provides us with a measure for the complexity of situations.

- $\mathrm{otp}(w) = 0$ iff $\{[v] \mid [v] < [w]\} = \emptyset$.
- $\mathrm{otp}(w) := \sup\{\mathrm{otp}(v) + 1 \mid [v] < [w]\}$, else.

For $M \subseteq \mathcal{F}$ let $M^\alpha := \{w \in M \mid \mathrm{otp}(w) \leq \alpha\}$. This measure of complexity is still quite rough. For example, all possibilities in $\mathcal{T}$ have order type $\mathrm{otp}(w) = 0$. We note the following fact, which follows by definition of otp.

Fact 3.1 *For $w \in \mathcal{F}$ we have:* $\forall v \in T(w)\,(\mathrm{otp}(v) = \mathrm{otp}(w) \Leftrightarrow w \in T(v))$. ⊟

Now, we can provide our final characterisation of $\mathcal{H}(\mathcal{T})$: It is the largest subclass of $\mathcal{F}$ such that for all $w \in \mathcal{H}(\mathcal{T}) \setminus \mathcal{T}$, and for all $X \in \mathrm{DP}$:

(**H$_1$**) $w \notin w(S) \cap w(H)$,
(**H$_2$**) $\forall v \in T(w)\,(w \notin v(X) \Rightarrow \forall u \in v(X)\,\mathrm{otp}(v) < \mathrm{otp}(w))$,
(**H$_3$**) $w(X) \subseteq \mathcal{H}(\mathcal{T})$.

The second axiom says that for all v in the transitive hull of w, if w is not an element of $v(X)$, then all possibilities in $v(X)$ have a complexity lower than the complexity of w. Following from Fact 3.1 (**H$_2$**) is equivalent to:

(**H$_2$**) $\forall v \in T(w)\,(w \notin v(X) \Rightarrow \forall u \in v(X)\,w \notin T(u))$.

4 Perspectives and Assertions

In this section we show how to derive the basic cases and the extended cases for asserting a sentence ψ using our perspectival operators. We concentrate on the *epistemic perspectives* of the dialogue participants and the role they play for assertions. The additional problems posed by constraints linked to speaker's goals are the topic of Section 5. We hope that the essentials of our theory become clearer if we separate the discussion of perspectives and speaker's intentions.

4.1 The Base Case Problem

We look again at the assertion (ψ) *It is snowing in the mountains* in (**1**).

In order to make his utterance, S should be convinced that it is really snowing in the mountains. If it in fact does but S has no evidence for ψ, then his assertion in (**1**) is not justified.

Suppose now it is snowing in the mountains, S is convinced of it, and H happens to know the truth of it too. But assume also that H is convinced that S can't know whether ψ. In this case the fact that the speaker asserts ψ contradicts his previous beliefs, and either he revises them or he has to assume that S is insincere or mistaken.

We make the assumption that *agents* are *justified* to perform an action iff they are convinced that it must be successful, and we make the assumption that hearers can make sense of an assertion if the fact that the act is performed by the speaker does not contradict the hearer's beliefs about the dialogue situation.

Assume that we have a condition γ which specifies the conditions of success for a certain dialogue act a. This means that the act can be performed successfully in all situations s where $s \models \gamma$. If a speaker S wants to perform a he has to be sure that γ really holds. Therefore, in a situation s where S performs a we must have $s \models \Box_S \gamma$. But this then becomes part of the information carried by the fact that a was performed. Hence, if the hearer recognises that a was performed by S, it should be the case that $s \models \Diamond_H(\gamma \wedge \Box_S\gamma)$. Otherwise the fact of S performing a will contradict H's beliefs.

But then, S has to be sure that this is the case. So we need in addition $s \models \Box_S \Diamond_H(\gamma \wedge \Box_S\gamma)$. We can go on with this way of reasoning and get an infinite number of new conditions for s. We can describe this observation by: Let Σ be the smallest set of formulas which contains γ and which is closed under: $p, q \in \Sigma$ then $p \wedge q, \Box_S p, \Diamond_H p \in \Sigma$. Then, we call the performance of dialogue act a in a situation s *mutually justified* iff $s \models \Sigma$. We claim that the situations in $\mathcal{T}$ in which a dialogue act is mutually justified are the *pragmatically basic* cases for this act. The following proposition gives us a simple criterion for deciding which situations in $\mathcal{T}$ support Σ.

Proposition 4.1 *With Σ defined as above and $s \in \mathcal{T}$ we get:*

$$s \models \Sigma \iff s \models \gamma \wedge \Box_S \gamma.$$
□

Proof: Let Σ be the smallest set containing γ and closed under $p, q \in \Sigma \Rightarrow p \wedge q, \Box_S p, \Diamond_H p \in \Sigma$. Let $M := \{s \in \mathcal{T} \mid s \models \gamma \wedge \Box_S \gamma\}$. It is clear that $s \models \Sigma$ implies $s \in M$. Hence, assume $s \in M$. Then $s \models \gamma$ and if $s \models p$ and $s \models q$ then, of course, $s \models p \wedge q$. Furthermore, $s \in \mathcal{T}$ implies $s \models p \Rightarrow s \models \Diamond_H p$. Hence, it remains to show that $s \models p$ implies $s \models \Box_S p$ for $p \in \Sigma$. If $p \equiv \gamma$, it follows by $s \in M$. Of course, we have $s \models \Box_S p \& s \models \Box_S q$ then $s \models \Box_S(p \wedge q)$. Hence, assume $p \equiv \Box_S q$ or $p \equiv \Diamond_H q$. The first case is clear due to introspection, and for the second part we have $s \models \Box_S q \Rightarrow s \models \Box_S \Diamond_H q$ because $s \in \mathcal{T}$. Therefore, we have $s \models \Box_S p$ for all $p \in \Sigma$. This finally proves that $s \models \Sigma$. ⊞

Let $M \subseteq \mathcal{T}$ be a class which represents some property of possibilities. Then, we can formulate our result as follows: Let $M^0 := M$, $M^{2n+1} := M^{2n} \cap \Box_S M^{2n}$, $M^{2n+2} := M^{2n+1} \cap \Diamond_H M^{2n+1}$, and

$$\nabla M := \bigcap_{n \in \mathbf{N}} M^n,$$

then

Lemma 4.2 *If $M \subseteq \mathcal{T}$, then $\nabla M = M \cap \Box_S M$.*
□

If a is the act where the speaker asserts a sentence ψ, then $\nabla[\psi]$ is the class of possibilities where the assertion is mutually justified. It is the class of basic cases.

4.2 The Derived Extended Uses

We now show how to derive extended uses. We proceed by recursion over the order type of possibilities. For this end, we define restricted versions of the operators. They produce only possibilities of a certain maximal complexity. Our aim is to construct in every step α exactly all possible extensions of complexity α. That we reached our aim will be proved in Lemma 4.3. For Example **(7)** it is essential that we can derive in later steps new extensions with the same complexity. In Lemma 4.3 we show that this can not happen. Let Q denote one of the operators $\square$ or $\Diamond$ and $\mathcal{H}_*(\mathcal{T}) := \mathcal{H}(\mathcal{T}) \setminus \mathcal{T}$.

$$Q_X^{\leq \alpha} M := Q_X M \cap \mathcal{H}_*^\alpha(\mathcal{T})$$
$$Q_X^{<\alpha} M := Q_X M \cap \{w \in \mathcal{H}_*^\alpha(\mathcal{T}) \mid \forall v \in w(X) \operatorname{otp}(v) < \alpha\}$$

The elements of $Q_X^{\leq \alpha} M$ all have maximal complexity α and an internal hierarchical structure. No element of $\mathcal{T}$ is an element of $Q_X^{\leq \alpha} M$. $Q_X^{<\alpha} M$ adds the restriction that all epistemic alternatives of participant X have a complexity *lower* than α. Hence, if $w \in Q_X^{<\alpha} M$ and $\operatorname{otp}(w) = \alpha$, then Fact 3.1 implies that $w \notin w(X)$.

With these operators we can define the operations which give us all extensions of a certain complexity. The following four operations correspond to the four possible intersections of classes, which we can derive using the four operators of Figure 2. For $\alpha > 0$ we define

$$\delta_1^\alpha M := \square_S^{\leq \alpha} M \cap \Diamond_H^{\leq \alpha} M$$
$$\delta_2^\alpha M := \square_S^{\leq \alpha} M \cap \Diamond_H^{\leq \alpha} \square_S^{<\alpha} M$$
$$\delta_3^\alpha M := \square_S^{\leq \alpha} \Diamond_H^{<\alpha} M \cap \Diamond_H^{\leq \alpha} M$$
$$\delta_4^\alpha M := \square_S^{\leq \alpha} \Diamond_H^{<\alpha} M \cap \Diamond_H^{\leq \alpha} \square_S^{<\alpha} M$$

For example, $w \in \delta_1^\alpha M$ means that the speaker is convinced that the order type of w is smaller than α and that it belongs to M, and that the hearer is also convinced that the order type of w is lower than α and that it might belong to M.

We define $\Delta_0(M) := M$, and $\Delta_{<\alpha}(M) := \bigcup_{\beta < \alpha} \Delta_\beta(M)$. For $\alpha > 0$ we set

$$\Delta_\alpha(M) := \Delta_{<\alpha}(M) \cup \bigcup_{i=1}^{4} \delta_i^\alpha \Delta_{<\alpha}(M).$$

We set $\Delta(M) := \bigcup_\alpha \Delta_\alpha(M)$. By definition of δ_i^α and $\Delta_{<\alpha}(M)$ it follows that $\Delta_\alpha(M) \subseteq \mathcal{H}^\alpha(\mathcal{T})$. The following lemma shows that we get for all α exactly all derived possible extensions of M of complexity α.

Lemma 4.3 *Let $M \subseteq \mathcal{T}$. For all ordinals α:*

$$\forall w \in \Delta_\alpha(M) \forall \beta < \alpha \; (w \in \Delta_\beta(M) \Leftrightarrow \operatorname{otp}(w) \leq \beta). \qquad \square$$

Proof: Let $w \in \Delta_\alpha(M)$ and $\beta < \alpha$. The direction from left to right is trivial. Assume that $\operatorname{otp}(w) = 0$. By definition of the operators $\square^\leq$, $\Diamond^\leq$ etc. it follows that $\Delta(M) \subseteq M \cup \mathcal{H}_*(\mathcal{T}) = M \cup (\mathcal{H}(\mathcal{T}) \setminus \mathcal{T})$. $w \in \mathcal{H}(\mathcal{T})$ and $\operatorname{otp}(w) = 0$ implies $w \in \mathcal{T}$ by Lemma A.1. Hence, $w \in \Delta(M) \& \operatorname{otp}(w) = 0$ implies $w \in M = \Delta_0(M)$. The next fact follows from (**H$_2$**) and from $w \in \mathcal{H}_*(\mathcal{T}) \Rightarrow \operatorname{otp}(w) > 0$:

Fact 4.4 $w \in \mathcal{H}_*(\mathcal{T})$ implies (1) $\forall v \in T(w)\, w \notin v(S) \cap v(H)$ and (2) $\forall v \in T(w)\, \neg \exists u, u'\, (u \in v(S)\, \&\, u' \in v(H)\, \&\, w \in T(u) \cap T(u'))$. □

So assume $\text{otp}(w) \leq \beta$, $\alpha = \beta+1$. Assume further that we have proved that $w \in \Delta_{\beta'}(M)$ for $\text{otp}(w) = \beta' < \beta$. Hence, let $\text{otp}(w) = \beta$. We consider only the case $w \in \delta_4^\alpha \Delta_{<\alpha}(M)$. Therefore, $w \in \square_S^{\leq \beta} \lozenge_H^{\leq \beta} \Delta_\beta(M) \cap \lozenge_H^{\leq \beta} \square_S^{\leq \beta} \Delta_\beta(M)$.

Let $v \in w(S)$, $\text{otp}(v) = \beta$. Then $v \in \lozenge_H^{\leq \beta} \Delta_\beta(M)$. Let $u \in v(H)$. Suppose $\text{otp}(u) = \beta$. Then $w \in T(u)$ by Fact 3.1, and therefore $w \in v(H)$ by (**H₂**). By (**H₂**) it follows also that $w \in w(S)$, and by introspection $w \in v(S)$. But then $w \in v(S) \cap v(H)$, in contradiction with Fact 4.4. Hence, for all $u \in v(H)\, \text{otp}(u) < \beta$. Therefore, $v \in \lozenge_H^{<\beta} \Delta_{<\beta}(M)$. As v was arbitrary, it follows that $w \in \square_S^{\leq \beta} \lozenge_H^{<\beta} \Delta_{<\beta}(M)$. If for all $v \in w(S)\, \text{otp}(v) < \beta$, then $w \in \square_S^{\leq \beta} \lozenge_H^{\leq \beta} \Delta_{<\beta}(M) \subseteq \square_S^{\leq \beta} \lozenge_H^{\leq \beta} \Delta_{<\beta}(M)$.

Let $v \in w(H) \cap \square_S^{\leq \beta} \Delta_\beta(M)$. Assume that $\text{otp}(v) = \beta$. Let $u \in v(S)$. Suppose $\text{otp}(u) = \beta$. Then $w \in T(u)$, and therefore $w \in v(S)$ by (**H₂**). But $v \in w(H) = v(H)$, therefore we also have $w \in v(H)$, in contradiction with Fact 4.4. Hence, for all $u \in v(S)\, \text{otp}(u) < \beta$. Hence, $v \in w(H) \cap \square_S^{<\beta} \Delta_{<\beta}(M)$, and therefore $w \in \lozenge_H^{\leq \beta} \square_S^{<\beta} \Delta_{<\beta}(M)$. If for all $v \in w(H)\, \text{otp}(v) < \beta$, then $w \in \lozenge_H^{<\beta} \square_S^{\leq \beta} \Delta_{<\beta}(M) \subseteq \lozenge_H^{\leq \beta} \square_S^{\leq \beta} \Delta_{<\beta}(M)$.

Hence, $w \in \square_S^{\leq \beta} \lozenge_H^{\leq \beta} \Delta_{<\beta}(M) \cap \lozenge_H^{\leq \beta} \square_S^{\leq \beta} \Delta_{<\beta}(M)$. Therefore $w \in \Delta_\beta(M)$.

Next, assume that α is a limit ordinal. Again, we consider only the case $w \in \delta_4^\alpha \Delta_{<\alpha}(M)$. Let $\text{otp}(w) = \beta < \alpha$. We suppose that we have proved that $v \in \Delta_{\beta'}(M)$ for $\text{otp}(v) = \beta' < \alpha$ and $v \in \Delta_{<\alpha}(M)$. This allows us to conclude that $w \in \square_S^{\leq \beta+1} \lozenge_H^{\leq \beta+1} \Delta_{<\beta+1}(M) \cap \lozenge_H^{\leq \beta+1} \square_S^{\leq \beta+1} \Delta_{<\beta+1}(M) = \delta_4^{\beta+1} \Delta_{<\beta+1}(M) \subseteq \Delta_{\beta+1}(M)$. By I.H. it follows that $w \in \Delta_\beta(M)$. ⊞

Remark 4.5 *Lemma 4.3 remains valid if we replace $\square_S$ in the definition of $\Delta(M)$ by an operator Q which has the form $Q\, M = \square_S(M \cap C)$ for some class $C \subseteq \mathcal{H}(\mathcal{T})$.* □

We can go through the proof of Lemma 4.3 and see that all inferences remain valid.

5 Speaker's Goals, Perspectives and Assertions

We introduce now explicit representations for the goals of the speaker. We model goals of a participant as a function mapping his epistemic possibilities into subsets of all possibilities, *i.e.*, this function tells us for each of his epistemic possibilities which situations are desirable for him. We have seen in Section 3.2 why we have to introduce speaker's goals. Example **(7)** shows that these goals together with some rationality constraints may lead to difficult problems. In case of extended uses we can avoid these problems by constructing the extensions relative $\mathcal{H}(\mathcal{T})$. In this class all possibilities have an internal hierarchical structure except for those possibilities which belong to $\mathcal{T}$. We will see that we can use essentially the same construction of extended uses for possibilities with speaker's goals. But the circular structures of $\mathcal{T}$ will lead to serious problems for the characterisation of basic cases.

Let $\mathcal{M}$ denote a class of models for the possible outer situations.

- A *possibility* is a triple $w = \langle s_w, \langle w(S), G_S^w \rangle, \langle w(H) \rangle \rangle$ such that
  - $s_w \in \mathcal{M}$,
  - $w(S)$ and $w(H)$ are information states,
  - G_S^w is a function with: (1) dom $G_S^w = w(S)$, and (2) $\forall v \in w(S)\, G_S^w(v)$ is an information state.
- An *information state* is a set of possibilities.

We denote the class of all possibilities with representations for the goals by $\mathcal{W}_G$. Transitivity of a class is defined in the same way as in Section 2: $M \subseteq \mathcal{W}_G$ is *transitive* iff $\forall w \in M\, \forall X \in \mathrm{DP}\, w(X) \subseteq M$. i.e., if $w \in M$, and if M is transitive, then the information states $G_S^w(v)$ don't need to be subsets of M.

For *introspectivity* we have to add a condition which guarantees that the real goals of the speaker and those of all his epistemic alternatives are the same. Let $\mathcal{I}_G$ denote the largest transitive subclass of $\mathcal{W}_G$ such that for all $X \in \mathrm{DP}$ for all $w \in \mathcal{I}_G$

- $\forall v \in w(X)\, w(X) = v(X)$,
- $\forall v \in w(S)\, G_S^w = G_S^v$.

We denote by $\dot{\mathcal{I}}_G$ the largest transitive subclass of $\mathcal{I}_G$ such that $\forall w \in \dot{\mathcal{I}}_G\, \forall X \in \mathrm{DP}\, w(X) \neq \emptyset$. We take $\mathcal{T}_G$ to be the largest transitive subclass of $\dot{\mathcal{I}}_G$ such that for all $w \in \mathcal{T}_G$

- $w \in w(S) \cap w(H)$
- $\forall v \in w(S)\, G_S^w(v) \subseteq \mathcal{T}_G$.

The second condition is a kind of *sincerity* condition, i.e., the speaker does not want to mislead the hearer.[9]

We modify our definition of *normal mutual update* for possibilities with speaker's goals: Let a be a function with domain $D(a)$, then a is a *normal mutual update* if (**NMU**) holds, and if

$$\forall w \in D(a)\, \forall v \in w(S) \cap D(a)\, G_S^{a(w)}(a(v)) = G_S^w(v).$$

i.e., we assume that an update with the information represented by $D(a)$ does not change speaker's goals.

The definitions of $T(w)$, otp and $\mathcal{H}(M)$ remain the same as in the previous sections except that $\mathcal{W}$ is replaced by $\mathcal{W}_G$. $\mathcal{F}_G$ denotes the subclass of $\dot{\mathcal{I}}_G$ where otp is defined.

In Section 3.2 we have introduced rationality constraints linked to speaker's goals. For an action a and a goal $G(w)$ the constraints looked as follows:

---

[9] We will see in Example (**11**) why we need such a condition.

$$(\mathbf{R_1}) \ w \notin G(w) \qquad (\mathbf{R_2}) \ a(w) \in G(w)$$

i.e., we should only perform an action if our goal is not yet achieved, and we should only choose actions which allow us to reach our goal.

We modify the $\Box$ operator in such a way that we check the rationality constraints at the time when we apply the operator. Let $M \subseteq \mathcal{F}_G$ be an arbitrary class:

$$[\mathrm{R}]_S M := \{w \in \dot{\mathcal{I}}_G \mid \forall v \in w(S) \, (v \in M \& v \notin G_S^v(v) \& a(v) \in G_S^v(v))\}$$
$$= \Box_S \left(M \cap \{v \in \dot{\mathcal{I}}_G \mid v \notin G_S^v(v) \& a(v) \in G_S^v(v)\} \right)$$

Remember that $G_S^w = G_S^v$ for all $v \in w(S)$. Therefore, we could also have written G_S^w instead of G_S^v in the definition of the operator. $[\mathrm{R}]_S M$ is the class of all possibilities where S is convinced that the real situation belongs to M, a can be performed successfully in order to reach his goal, and where this goal is not already reached. The axioms ($\mathbf{R_1}$) and ($\mathbf{R_2}$) hold by definition of the operator for all epistemic alternatives of S. It follows that $\Diamond_H [\mathrm{R}]_S M$ denotes the class of all possibilities where H thinks it is possible that S can choose a in order to reach his goal.

5.1 The Base Case Problem

In Section 4.1 we argued that the speaker should be convinced that his assertion is successful, and the hearer should know that this can be the case. Then the speaker should be convinced that both conditions hold, and the hearer should again know that this is possible. We have argued that, in principle, we must impose these conditions again and again. If M denotes the class where some sentence ψ is true, then we denote the class where its assertion is *mutually justified* as ∇M. For possibilities without speaker's goals we could provide a simple characterisation of ∇M. In this section we will see that this is not possible if we add goals.

We see that the condition for the hearer does not impose a real restriction because $M \subseteq \mathcal{T}_G$ implies $M \cap \Diamond_H M = M$. Hence, $M \cap [\mathrm{R}]_S M \cap \Diamond_H M = M \cap [\mathrm{R}]_S M$. Let a_M denote the normal mutual update determined by M. With $w \in \mathcal{T}_G \Rightarrow w \in w(S)$ and introspection we find:

$$M \cap [\mathrm{R}]_S(M \cap [\mathrm{R}]_S M) =$$
$$= M \cap [\mathrm{R}]_S \{w \in M \mid w(S) \subseteq M \& \forall v \in w(S)(v \notin G_S^w(v) \& a_M(v) \in G_S^w(v))\}$$
$$= \{w \in M \mid w(S) \subseteq M \& \forall v \in w(S)(v \notin G_S^w(v) \& a_M(v) \in G_S^w(v))\}$$
$$= M \cap [\mathrm{R}]_S M.$$

If we adopt the definition from Section 4.1 and define ∇M by $M^1 := M \cap [\mathrm{R}]_S M$, $M^{n+1} := M^n \cap [\mathrm{R}]_S M^n \cap \Diamond_H M^n$, $\nabla M := \bigcap_{n \in \mathbb{N}} M^n$, then the considerations above seem to show that we can again characterise ∇M as $M \cap [\mathrm{R}]_S M$. But the following example shows that this is not correct:

(8) Helga calls up her son Stephan early on Sunday morning and asks him whether he wants to visit her. They both know that Stephan would have checked the weather at this time only if he needed an excuse for not accepting the expected invitation. It happens that he knows that it is snowing, and he does not want to let her know that he would prefer to stay at home this day. Should he tell his mother that he can't come because *it is snowing in the mountains*? ⌑

In Example **(8)** it is true that (ψ) *it is snowing in the mountains*, and we assume that, if Helga learns this fact, she will excuse her son for not visiting her, and that she can't know whether Stephan likes to visit her or not. Hence, if M denotes the set of all $w \in \mathcal{T}_G$ where *it is snowing in the mountains*, then the situation is an element of $M \cap [\mathrm{R}]_S M$, and we would predict that Stephan will be successful if he tells her that ψ. But of course this implies that Helga also learns that Stephan knows that ψ, hence she can conclude that he does not like to visit her, which is an undesired result for Stephan. This means that Stefan should check in the second step whether the mutual update with $M \cap [\mathrm{R}]_S M$ also leads to a desired situation. The problem is that the above definition of ∇M does only capture the fact that the hearer learns ψ but not that she also learns that the speaker knows that ψ and that it is reasonable for him to assert ψ, and that he knows that she knows this, *etc.*

In general, we have to find a fixed–point of the update operations. This is a non–trivial task. First of all, we have to find the correct fixed–point condition. In order to capture Example **(8)**, we have to search for sets M which are fixed–points of the following operator. Let $M \subseteq \mathcal{T}_G$:

$$J M := \{w \in M \mid w(S) \subseteq M \& \forall v \in w(S)(v \notin G_S^w(v) \& a_M(v) \in G_S^w(v))\}$$

Let $M_0 := M$ and $M_{n+1} := J M_n$. Then, M_1 is the set of worlds where the rationality conditions for asserting M are satisfied. It follows especially that $v \in M_1$ implies that mutually updating with M leads to a desired outcome. M_1 may be a proper subset of M, hence, mutually learning that M_1 may lead to a different outcome than simply mutually learning that M. Hence, we have to check in the next step whether mutually learning M_1 is desired by the speaker. In general, in step $n + 1$ we check whether mutually learning M_n is desirable for the speaker.

In this paper, we use a strong assumption about speaker's goals which guarantees that a fixed–point can easily be constructed. The assumption states: if the speaker can reach his goal by a mutual update with some information M, then he can also reach it by updating with any stronger information $N \subseteq M$. Hence, in the following, we consider only sets $M \subseteq \mathcal{T}_G$ such that for all $w \in M$ and $v \in w(S)$:

(R$_3$) $\forall N \subseteq M : a_M(v) \in G_S^w(v) \& v(S) \subseteq N \Rightarrow a_N(v) \in G_S^w(v)$

With the foregoing, we can easily show the following lemma.

Lemma 5.1 *Let $M \subseteq \mathcal{T}_G$ be such that ($\mathbf{R_3}$) holds for all $v \in M$. Then, $\nabla M := M \cap [R]_S M$ is a fixed–point the J operator.* □

5.2 The Derived Extended Uses

In this section we show how to derive extended uses of assertion if we add rationality constraints linked to speaker's goals to our considerations. In principle we proceed in the same way as in Section 4.2 by recursively defining extensions $\Delta_\alpha(\mathcal{B})$ of some base set $\mathcal{B}$. If we know $\Delta_\alpha(\mathcal{B})$ and want to decide whether some w belongs to $\Delta_{\alpha+1}(\mathcal{B})$, then we have to check wether the update with the information carried by the assertion leads to a desirable state. But this means that we have to know $\Delta_{\alpha+1}(\mathcal{B})$ in order to check the rationality principles. As a consequence, we cannot directly define $\Delta_{\alpha+1}(\mathcal{B})$ by using the operators linked to perspectives. Therefore we proceed as follows: Given some set $M \subseteq \mathcal{T}_G$, we provide an axiomatic characterisation of possible extensions $\Delta_\alpha(M)$ of M and then show that there exists exactly one sequence $(\Delta_\alpha(M))_{\alpha \in \mathbf{On}}$ which satisfies the following axioms ($\mathbf{D_1}$) and ($\mathbf{D_2}$). Let $(\Delta_\alpha(M))_{\alpha \in \mathbf{On}}$ be a sequence of subclasses of $\mathcal{H}(\mathcal{T})$. The first axiom ($\mathbf{D_1}$) states that for each α the elements of $\Delta_\alpha(M)$ with order type $\beta \leq \alpha$ are already elements of $\Delta_\beta(M)$.

($\mathbf{D_1}$) $\forall \beta \leq \alpha \ \Delta_\alpha(M) \cap \mathcal{H}^\beta(\mathcal{T}) = \Delta_\beta(M)$.

Note that ($\mathbf{D_1}$) implies $\Delta_0(M) = M$ and $\Delta_\alpha(M) \subseteq \mathcal{H}^\alpha(\mathcal{T})$.

Let a_α be the normal mutual update determined by $\Delta_\alpha(M)$, and a_Δ the normal mutual update determined by $\Delta(M) := \bigcup_\alpha \Delta_\alpha(M)$. ($\mathbf{D_1}$) implies $a_\alpha = a_\Delta|_{\mathcal{H}^\alpha(\mathcal{T})}$, and especially it follows $w \in \Delta_\alpha(M) \Rightarrow a_\alpha(w) = a_\Delta(w)$. We define the operator [R] which combines the $\square$ operator and the checking of rationality axioms ($\mathbf{R_1}$), ($\mathbf{R_2}$) relative to a_Δ:

$$[R]_S N := \{w \in \dot{\mathcal{I}}_G \mid \forall v \in w(S)\, (v \in N \,\&\, v \notin G_S^v(v) \,\&\, a_\Delta(v) \in G_S^v(v))\}.$$

The next axiom demands that the new elements of $\Delta_\alpha(M)$ are those which can be derived by application of the restricted versions of perspectival operators motivated in Section 3.

($\mathbf{D_2}$) Let $w \in \mathcal{H}^\alpha(\mathcal{T})$ and $\Delta_{<\alpha}(M) := \bigcup_{\beta < \alpha} \Delta_\beta(M)$. Then $w \in \Delta_\alpha(M)$, iff $w \in \Delta_{<\alpha}(M)$ or there exists $N \subseteq \Delta_{<\alpha}(M)$ such that w is an element of one of the following sets:

$$\delta_1^\alpha N := [R]_S^{\leq \alpha} N \cap \diamond_H^{\leq \alpha} N; \qquad \delta_2^\alpha N := [R]_S^{\leq \alpha} N \cap \diamond_H^{\leq \alpha} [R]_S^{\leq \alpha} N$$

$$\delta_3^\alpha N := [R]_S^{\leq \alpha} \diamond_H^{\leq \alpha} N \cap \diamond_H^{\leq \alpha} N; \qquad \delta_4^\alpha N := [R]_S^{\leq \alpha} \diamond_H^{\leq \alpha} N \cap \diamond_H^{\leq \alpha} [R]_S^{\leq \alpha} N$$

We call $(\Delta_\alpha)_{\alpha \in \mathbf{On}}$ a *possible derived extension* of M if the axioms ($\mathbf{D_1}$) and ($\mathbf{D_2}$) hold.

The condition of ($\mathbf{D_2}$) does not allow for a direct definition of $\Delta_\alpha(M)$ because the definition of the [R] operator presupposes that $\Delta_\alpha(M)$ is already defined. We show that for every $M \subseteq \mathcal{T}_G$ there exists a unique possible derived extension $\Delta(M)$, see Theorem 5.3. We define an extension $\Delta(M)$ and a_Δ simultaneously by recursion over α.

Assume that $\Delta_\beta(M)$ and a_β are defined for $\beta < \alpha$ and that the restricted versions of (**D$_1$**) and (**D$_2$**) hold for $(\Delta_\beta(M))_{\beta<\alpha}$. We show that this already implies that $[R]_{\overline{S}}^{\leq\alpha}N$ and $[R]_{\overline{S}}^{\leq\alpha}\Diamond_H^{\leq\alpha}N$ are uniquely defined for $N \subseteq \Delta_{<\alpha}(M)$. First we see that by definition $[R]_{\overline{S}}^{\leq\alpha}N$ must be equal to:

$$\{w \in \mathcal{H}^\alpha(\mathcal{T}) \mid w(S) \subseteq N \& \forall v \in w(S)(a_\Delta(v) \in G^v(v) \& v \notin G^v(v))\}$$

If we write $a_{<\alpha}$ for the normal mutual update determined by $\Delta_{<\alpha}(M)$, then (**D$_1$**) implies that this is equal to:

$$\{w \in \mathcal{H}^\alpha(\mathcal{T}) \mid w(S) \subseteq N \& \forall v \in w(S)(a_{<\alpha}(v) \in G^v(v) \& v \notin G^v(v))\}.$$

But $a_{<\alpha}$ is given by I.H., hence, $[R]_{\overline{S}}^{\leq\alpha}N$ is uniquely defined.

Next we consider $[R]_{\overline{S}}^{\leq\alpha}\Diamond_H^{\leq\alpha}N$. By definition it must be equal to:

$$\{w \in \mathcal{H}^\alpha(\mathcal{T}) \mid w(S) \subseteq \Diamond_H^{\leq\alpha}N \& \forall v \in w(S)(a_\Delta(v) \in G^v(v) \& v \notin G^v(v))\}$$

Hence, due to introspection, $w \in [R]_{\overline{S}}^{\leq\alpha}\Diamond_H^{\leq\alpha}N$ implies $w(S) \subseteq [R]_{\overline{S}}^{\leq\alpha}\Diamond_H^{\leq\alpha}N$. This implies that $a_\Delta(w(S)) = \{a_\Delta(v) \mid v \in w(S)\}$. Hence, it must hold that

(*) $\quad a_\Delta(w) = \langle s_w, \langle \{a_\Delta(v) \mid v \in w(S)\}, G^{a_\Delta(w)} \rangle, \langle a_\Delta(w(H)) \rangle \rangle.$

We can assume that $w \in w(S)$, hence $w(H) \subseteq \Diamond_H^{\leq\alpha}N$. As $a_\Delta|_{w(H)}$ is defined by I.H., it follows that $a_\Delta(w)$ is uniquely defined by this equation[10]. Hence, we can define $[R]_{\overline{S}}^{\leq\alpha}\Diamond_H^{\leq\alpha}N$. This shows that $\Delta_\alpha(M)$ exists and is uniquely determined by (**D$_2$**) and I.H. We have to check that (**D$_1$**) and (**D$_2$**) hold also for the sequence $(\Delta_\alpha(M))_{\beta\leq\alpha}$. First, we show (**D$_1$**), i.e., we show that we get in step α exactly all new situations of complexity α:

Lemma 5.2 $\forall \gamma \leq \beta \leq \alpha \; \Delta_\alpha(M)^\gamma = \Delta_\beta(M)^\gamma.$ □

But this follows from Remark 4.5 because of:

$$[R]_S N = \square_S \left(N \cap \{v \in \dot{\mathcal{I}}_G \mid v \notin G^v_S(v) \& a_\Delta(v) \in G^v_S(v)\} \right)$$

This shows that (**D$_1$**) holds. But this means that $a_\alpha|_{\mathcal{H}^{<\alpha}(\mathcal{T})} = a_{<\alpha}$. Therefore, it follows that a_α is also a solution for (*). This implies that (**D$_2$**) holds. Hence, this construction defines recursively a unique extension a_Δ simultaneously with $\Delta(M)$. We summarise the result as:

Theorem 5.3 *Let $M \subseteq \mathcal{T}_G$. Then there exists a unique possible extension $\Delta(M)$.* □

---

[10] The last equation defines a *system of equations*, where the $a_\Delta(v)$ for $v \in T(w)$ with $\operatorname{otp}(v) \geq \alpha$ denote new *urelements* which function as unknown parameters. (*AFA*) set theory guarantees that it has a unique solution for *all* w. Then, the set of new possibilities in $[R]_{\overline{S}}^{\leq\alpha}\Diamond_H^{\leq\alpha}N$ is equal to the set of all $w \in \mathcal{H}^\alpha(\mathcal{T})$ such that (1) $w(H) \subseteq \Diamond_H^{\leq\alpha}N$ and such that (2) the solution a_Δ of (*) satisfies $\forall v \in w(S)(a_\Delta(v) \in G^v(v) \& v \notin G^v_S(v))$.

6 Applications

We apply our theory to examples introduced in the previous sections. A possibility w is a triple $\langle s_w, \langle w(S), G_S^w \rangle, \langle w(H) \rangle \rangle$ where s_w is a model for the *outer* situation. As we are only interested in the truth or falsity of the sentence (ψ) '*It is snowing in the mountains*', we use the respective formula to denote this model. We denote by $[\psi]$ the set of all possibilities in $\mathcal{T}_G$ where ψ is true. We write a_M for the normal mutual update determined by some M. We can summarise the results of the last sections as follows:

- In a basic case an assertion *that ψ* must be mutually justified. We denote the class of all these basic situations by $\mathcal{B} = \nabla[\psi]$.
- The maximal class where the speaker can reasonably assert that ψ, and where the hearer can interpret this assertion, is given by $\Delta(\mathcal{B})$.
- The update effect is given by $a_{\Delta(\mathcal{B})}$, the normal mutual update determined by $\Delta(\mathcal{B})$.

Theorem 5.3 guarantees that $\Delta(\mathcal{B})$ and $a_{\Delta(\mathcal{B})}$ are always defined.

The basic case is exemplified by Example **(1)**. We can represent the utterance situation w_1 by the following equations[11]:

$$w_1 = \langle \psi, \langle \{w_1\}, G_S^{w_1} \rangle, \langle \{w_1, v_1\} \rangle \rangle$$
$$v_1 = \langle \neg\psi, \langle \{v_1\}, G_S^{v_1} \rangle, \langle \{w_1, v_1\} \rangle \rangle,$$

i.e., it is the case that ψ, H does not know it but knows that S knows whether ψ. The intended resulting state is described by the equation

$$s_1 = \langle \psi, \langle \{s_1\}, G_S^{s_1} \rangle, \langle \{s_1\} \rangle \rangle,$$

where $G_S^{s_1}(s_1) = G_S^{w_1}(w_1)$. Hence, we assume that $\{s_1\} = G_S^{w_1}(w_1)$, and $\{t_1\} = G_S^{v_1}(v_1)$, where $t_1 = \langle \neg\psi, \langle \{t_1\}, G_S^{t_1} \rangle, \langle \{t_1\} \rangle \rangle$, i.e., if ψ holds, then S wants them mutually to know that ψ, and if $\neg\psi$ holds, then he wants them mutually to know that $\neg\psi$ holds. Clearly, $w_1, v_1, s_1 \in \mathcal{T}_G$. We find that $w_1 \notin G_S^{w_1}(w_1)$, and $s_1 \in G_S^{w_1}(w_1)$. Hence, $w_1 \in [R]_S\{w_1\}$, and therefore by Lemma 5.1: $w_1 \in \nabla\mathcal{B}$.

For a proof of $a_{\mathcal{B}}(w_1) = s_1$ we would need some additional techniques from (AFA) set theory. The general properties of normal mutual updates imply

$$a_{\mathcal{B}}(w_1) = \langle \psi, \langle \{a_{\mathcal{B}}(w_1)\}, G_S^{a_{\mathcal{B}}(w_1)} \rangle, \langle \{a_{\mathcal{B}}(w_1)\} \rangle \rangle.$$

But this equation is structurally identical with the equation for s_1. Using the solution lemma, Theorem 2.1, it is provable that the two equations have the same solution.

---

[11] Of course, this interpretation is not fully justified by the way the example was stated. There are a lot of dialogue situations where this can be part of the description. We don't want to explain how we arrive at such a strong reading, but merely, given the reading, why the utterance of ψ is reasonable for the speaker and can be interpreted by the hearer.

Hence, the theory predicts that it is reasonable for S to say that ψ, and that it will be successful.

In Example **(2)** the beliefs and goals are the same as in **(1)**:

$$w_2 = \langle \neg\psi, \langle\{w_1\}, G_S^{w_1}\rangle, \langle\{w_1, v_1\}\rangle\rangle.$$

Hence, it is an element of $\square_S^{\leq 1}\mathcal{B} \cap \diamondsuit_H^{\leq 1}\mathcal{B} = \delta_1^1\mathcal{B} \subseteq \Delta_1(\mathcal{B})$. Hence it is reasonable for S to say that ψ, and the hearer will interpret his utterance in the same way as in situation **(1)**. Next, we consider Example **(3)**.

$$w_3 = \langle \neg\psi, \langle\{w_1\}, G_S^{w_1}\rangle, \langle\{w_3\}\rangle\rangle.$$

Here, the information of the speaker is the same as in the basic situation w_1, and as in w_2. But this time, the hearer knows that it is not snowing in the mountains ($\neg\psi$), and she is aware of the entire dialogue situation. w_3 is an element of $\square_S^{\leq 1}\{w_1\} \cap \square_H^{\leq 1}\square_S^{\leq 1}\{w_1\} \subseteq \delta_2^1\mathcal{B} \subseteq \Delta_1(\mathcal{B})$. The update of w_2 and w_3 with $a_{\Delta_1(\mathcal{B})}$ leads to s_2 and s_3, where

$$s_2 = \langle \neg\psi, \langle\{s_1\}, G_S^{s_1}\rangle, \langle\{s_1\}\rangle\rangle,$$
$$s_3 = \langle \neg\psi, \langle\{s_1\}, G_S^{s_1}\rangle, \langle\{s_3\}\rangle\rangle.$$

In the next example, Example **(9)**, S is no longer sincere. He is lying successfully. It is a case in which the belief state of the hearer is the same as in the basic situation, and where the speaker knows this but knows also that ψ is not true. The speaker can exploit this situation and deceive the hearer.

(9) Helga calls up her son Stephan and asks him whether he wants to visit her in Munich. Stephan, who has absolutely no inclination to drive to Munich this day, answers: "It is snowing in the mountains."

Now, the hearer updates as in the basic case, and the speaker updates only his representation of the hearer's beliefs.

$$w_4 = \langle \neg\psi, \langle\{w_4\}, G_S^{w_4}\rangle, \langle\{w_1, v_1\}\rangle\rangle,$$

with $G_S^{w_4}(w_4) = \{s_4\}$, where

$$s_4 = \langle \neg\psi, \langle\{s_4\}, G_S^{s_4}\rangle, \langle\{s_1\}\rangle\rangle,$$

and $G_S^{s_4}(s_4) = \{s_4\}$. w_4 is an element of $\square_S^{\leq 1}\diamondsuit_H^{\leq 1}\{w_1\} \cap \diamondsuit_H^{\leq 1}\{w_1\} \subseteq \delta_3^1\mathcal{B} \subseteq \Delta_1(\mathcal{B})$. We can see that $a_{\Delta_1(\mathcal{B})}(w_4) = s_4$. Hence, the speaker should be able to successfully mislead the hearer.

Example **(4)** receives the following representation:

$$w_5 = \langle \neg\psi, \langle\{w_4\}, G_S^{w_4}\rangle, \langle\{w_5\}\rangle\rangle.$$

i.e., the speaker remains in the same situation as in Example **(9)**, but now the hearer knows that the speaker wants to mislead him. w_5 is an element of $\square_S^{\leq 1}\diamondsuit_H^{\leq 1}\{w_1\} \cap \diamondsuit_H^{\leq 2}\square_S^{\leq 2}\{w_4\} \subseteq \delta_4^2\{w_4\} \subseteq \Delta_2(\mathcal{B})$. We can see that

$a_{\Delta_2(\mathcal{B})}(w_5) = s_5$ with $s_5 = \langle \neg \psi, \langle \{s_4\}, G_S^{s_4} \rangle, \langle \{s_5\} \rangle \rangle$. Hence, the speaker thinks that he has successfully misled the hearer, and the hearer knows this.

The next example is a case where the hearer *suspects* that the speaker might lie.

(10) Helga calls up her son Stephan. She knows that he does not like to visit her, hence she suspects that he will not be honest. But she knows also that Stephan can't know this. She asks him whether he wants to visit her, and Stephan replies that *it is snowing in the mountains*. And, indeed, it was not a lie. □

$$w_7 = \langle \psi, \langle \{w_1\}, G_S^{w_1} \rangle, \langle \{w_7, u_7, v_7\} \rangle \rangle$$
$$u_7 = \langle \neg \psi, \langle \{v_1\}, G_S^{v_1} \rangle, \langle \{w_7, u_7, v_7\} \rangle \rangle$$
$$v_7 = \langle \neg \psi, \langle \{w_4\}, G_S^{w_4} \rangle, \langle \{w_7, u_7, v_7\} \rangle \rangle.$$

This is an element of $\Delta_2(\mathcal{B})$, and the update with $a_{\Delta_2(\mathcal{B})}$ results in s_7:

$$s_7 = \langle \psi, \langle \{s_1\}, G_S^{s_1} \rangle, \langle \{s_7, t_7\} \rangle \rangle$$
$$t_7 = \langle \neg \psi, \langle \{s_4\}, G_S^{s_4} \rangle, \langle \{s_7, t_7\} \rangle \rangle.$$

i.e., Stephan believes that they now mutually believe that it is snowing in the mountains. His mother knows this to be the case if it were really snowing, and she believes that, if it were not snowing, he would think that he could deceive her successfully.

The following example shows that we need a *sincerity* condition for our basic situations. In (11) it is common knowledge that the hearer thinks that the speaker might lie.

(11) Helga calls up her son Stephan. The last time she invited him, he pretended that he could not come because of the bad weather. They both know that she will be suspicious this time, if he replies to her question that *it is snowing in the mountains*. □

$$w_8 = \langle \psi, \langle \{w_8\}, G_S^{w_8} \rangle, \langle \{w_8, v_8\} \rangle \rangle$$
$$v_8 = \langle \neg \psi, \langle \{v_8\}, G_S^{v_8} \rangle, \langle \{w_8, v_8\} \rangle \rangle,$$

where $G_S^{w_8}(w_8) = \{s_1\}$ and $G_S^{v_8}(v_8) = \{s_4\}$. Clearly, $otp(w_8) = 0$. But w_8 is not an element of $\mathcal{T}_G$ because $G_S^{v_8}(v_8) \not\subseteq \mathcal{T}_G$. Hence, $w_8 \notin \mathcal{T}_G \supseteq \nabla[\psi] = \mathcal{B}$. Therefore, it can't be an element of $\Delta(\mathcal{B})$. Without sincerity condition, *i.e.*, without the condition $G_S^w(v) \subseteq \mathcal{T}_G$ for $w \in \mathcal{T}G$ and $v \in w(S)$, w_8 would be an element of $\mathcal{T}_G$, and the general theory would predict that the hearer would update his belief state with the information that ψ. Intuitively, Stephan has to make clear that he is sincere to make his mother believe him. *i.e.*, they mutually have to redefine $G_S^{v_8}(v_8)$ as t_1, but this results in w_1 and then, indeed, the assertion of ψ will result in s_1.

7 Conclusion

We searched for a characterisation of dialogue situations where a felicitous use of an assertion is possible. Felicity is taken here in the sense that (1) the speaker is convinced that the update triggered by the assertion leads to a desirable situation, and (2) the fact that the sentence was asserted does not contradict the beliefs of the interpreter. We distinguished between the use of assertions in *ideal* and *non–ideal* situations, where in ideal situations it holds that it is common knowledge that (1) everybody believes the real situation to be possible (*has knowledge*), and (2) the speaker does not want to mislead the hearer. We started with the felicity conditions and update effects of assertions as they are defined by pure semantics, *i.e.*, we assume that a sentence with semantic content ψ can be asserted exactly iff ψ is true in the actual situation. We described the effect of an assertion by a normal mutual update with its semantic content. AFA set theory was an indispensable tool for developing our theory.

The pragmatically *basic* cases build the subclass $\mathcal{B}$ where the pragmatic constraints linked to perspectives and intentions hold. We saw that the implicit circularity of these constraints makes the task of finding $\mathcal{B}$ a non–trivial one.

We have argued that perspectives play an essential role in explaining extended uses in non–ideal situations. The idea was to start with a characterisation of basic situations, and then to derive new uses from ideal situations by systematic application of operators which reflect the way partiality of information can give rise to uses in new situations.

The theory presented in this paper evolved out of our research concerning the felicity conditions for the referential use of definite descriptions (Benz, 1999). There we used the same ideas to explain the referential use in situations with arbitrarily complex belief structures. The present paper contains a much improved description of the underlying mathematical structures. In (Benz, 2000) we outlined on an informal level how the two papers combine to form a single theory of perspectives. Assertions are examples of speech acts, and speech acts are examples of goal–directed linguistic actions. If a linguistic act can be characterised by its preconditions, the speaker's goals and if its effects can be described by mutual learning, then our approach should be applicable and explain how this linguistic act can have extended uses in defective dialogue situations. If successful, this theory greatly simplifies the analysis of linguistic acts. To find a characterisation of linguistic acts in ideal situations is normally a less complicated task. The theory of perspectival derivations then automatically predicts their use and effect in defective situations.

References

Peter Aczel. *Non-well-founded sets*. Center for the study of language and information, Stanford CA, 1988.

Alexandru Balta, Lawrence Moss, and Slawomir Solecki. The logic of public announcements, common knowledge and private suspicion. Technical report, Centrum for wiskunde en informatica, Amsterdam, 1999.

Alexandru Baltag. A logic of epistemic actions. In *Proccedings of the ESSLI workshop on foundations and applications of collective agent based systems*, Utrecht, 1999. Available from http://www.let.uu.nl/esslli/Courses/hoek.html.

Alexandru Baltag, Hans van Ditmarsch, and Lawrence Moss. Epistemic logic and information update. In *Handbook on the philosophy of information*. Elsevier Science Publishers, to appear.

Jon Barwise and Lawrence Moss. *Vicious circles*. Center for the study of language and information, 1996.

Anton Benz. Perspectives and the referential use of definite descriptions. ms., Berlin, 1999.

Anton Benz. Perspectives and derived extensions of dialogue acts. *LDV-Forum*, 17:115–136, 2000. Proceedings of the Communicating Agents workshop, Bonn.

Ronald Fagin, Joseph Y. Halpern, Yoram Moses, and Y. Vardi Moshe. *Reasoning about knowledge*. MIT Press, Cambridge MA, 1995.

Abraham A. Fraenkel and Yehoshua Bar-Hillel. *Foundations of set theory*. North-Holland Publishing Company, 1958.

Jelle Gerbrandy. *Bisimulations on planet Kripke*. PhD thesis, Institute for logic, language and computation, Universiteit van Amsterdam, 1998.

Jelle Gerbrandy and Willem Groeneveld. Reasoning about information change. *Journal of logic, language and information*, 6:147–169, 1997.

Willem Groeneveld. *Logical investigations into dynamic semantics*. PhD thesis, Institute for logic, language and computation, Universiteit van Amsterdam, 1995.

Jaakko Hintikka. *Knowledge and belief*. Cornell University Press, Ithaca and London, 1962.

Jan O. M. Jaspars. *Calculi for constructive communication. A study of the dynamic of partial sets*. PhD thesis, ITK, Katholieke Universitet Brabant, 1994.

Henk Zeevat. The common ground as a dialogue parameter. In Anton Benz and Gerhard Jäger, editors, *Proceedings of the Munich workshop on formal semantics and pragmatics of dialogue MunDial'97*, pages 195–214, Munich, 1997. Centrum für Informations- und Sprachverarbeitung.

A The Structure of $\mathcal{H}(\mathcal{T})$

In this section we prove the claims we have made about $\mathcal{H}(\mathcal{T})$. We first show that "$(\mathbf{H_1})$–$(\mathbf{H_3})$ + $\mathcal{H}(\mathcal{T}) \subseteq \mathcal{F}$" defines the same class as "$(\mathbf{H_{1'}})$–$(\mathbf{H_{3'}})$ + $\mathcal{H}(\mathcal{T}) \subseteq \dot{\mathcal{I}}$". In the second part we show that we can construct $\mathcal{H}(\mathcal{T})$ by an iterated application of a combination of perspectival operators. The proofs in the subsequent sections do only make use of "$(\mathbf{H_1})$–$(\mathbf{H_3}) + \mathcal{H}(\mathcal{T}) \subseteq \mathcal{F}$"[12]. The following lemma shows that this implies that (1) the complexity of the possibilities $w \in \mathcal{H}(\mathcal{T}) \setminus \mathcal{T}$ is greater than at least one participant believes it to be, (2) all situations with complexity 0 must be elements of $\mathcal{T}$, and (3) $(\mathbf{H_{3'}})$ holds.

Lemma A.1 *Let $w \in \mathcal{H}(\mathcal{T})$. Then*

1. *If $w \notin \mathcal{T}$, then $\exists X \in \mathrm{DP}$ such that $\forall v \in w(X) \mathrm{otp}(v) < \mathrm{otp}(w)$.*
2. *$\mathrm{otp}(w) = 0 \vee w \in w(S) \cap w(H) \Rightarrow w \in \mathcal{T}$.*
3. *Let $X \neq Y$. If p is a function with $p(0) = w$, $p(2n+1) \in p(2n)(X)$, and $p(2n+2) \in p(2n+1)(Y)$, then there is an $n \in \mathbf{N}$ with $p(n) \in \mathcal{T}$.* □

Proof: (1) By $(\mathbf{H_1})$ there exists $X \in \mathrm{DP}$ such that $w \notin w(X)$. With $(\mathbf{H_2})$ it follows for all $v \in w(X)$ that $w \notin T(v)$, which implies $\mathrm{otp}(w) \geq \mathrm{otp}(v) + 1$.

(2) If $w \in w(S) \cap w(H)$, it immediately follows from $(\mathbf{H_1})$ that $w \in \mathcal{T}$. If $w \notin \mathcal{T}$, it follows from 1. that $\mathrm{otp}(w) \geq 1$.

(3) It follows from 1. and $(\mathbf{H_2})$ that $\mathrm{otp}(p(n+2)) < \mathrm{otp}(p(n))$. Suppose that for all $n \in \mathbf{N}$ $p(n) \notin \mathcal{T}$. Then there exists an n with $\mathrm{otp}(p(n)) = 0$. But then, $p(n) \in \mathcal{T}$, which contradicts the assumption. ⊞

As $(\mathbf{H_{1'}})$ is the same as $(\mathbf{H_1})$ and $(\mathbf{H_{3'}})$ implies $(\mathbf{H_{2'}})$, it follows that $(\mathbf{H_1})$–$(\mathbf{H_3}) + \mathcal{H}(\mathcal{T}) \subseteq \mathcal{F}$ imply $(\mathbf{H_{1'}})$–$(\mathbf{H_{3'}})$. Clearly, $(\mathbf{H_{3'}})$ implies for $w \in \dot{\mathcal{I}}$ that $w \in \mathcal{F}$. We need the following fact:

Fact A.2 *Let $X, Y \in \mathrm{DP}$, $X \neq Y$ and $w, v \in \mathcal{F}$. If $v \in w(X)$, then $T(v) = w(X) \cup \bigcup_{u \in v(X)} \bigcup_{u' \in u(Y)} T(u') =: T'(v)$.* □

Clearly $T'(v) \subseteq T(v)$. As $T'(v)$ is transitive and $\{v\} \subseteq T'(v)$, it follows from definition of $T(v)$ that $T(v) \subseteq T'(v)$. ⊞

If $(\mathbf{H_2})$ does not hold for $w \in \mathcal{F}$, then there is a $v \in T(w)$ such that for some X $w \notin v(X) \& \exists u \in v(X)$ $w \in T(u)$. Fact A.2 implies that $w \in u'(Y)$ for some $u' \in v(X)$, $Y \neq X$. But this means that we can construct a sequence $(v_n)_{n \in \mathbf{N}}$ which violates $(\mathbf{H_{2'}})$.

We can now provide a construction of $\mathcal{H}(\mathcal{T})$ which shows that we can derive this class by systematic application of perspectival operators. We will even show that we can apply them in such a way that we get in each step α of the recursion exactly all elements of $\mathcal{H}(\mathcal{T})$ with complexity α. We therefore introduce restricted versions of our operators:

- $\Box_X^{\leq \alpha} M := \Box_X M \cap \mathcal{F}^\alpha$,
- $\Box_X^{<\alpha} M := \{w \in \mathcal{F}^\alpha \mid w(X) \subseteq M \& \forall v \in w(X) \, \mathrm{otp}(v) < \alpha\}$.

This gives us more control over the complexity of derived candidates. We can then construct our class of candidates as suggested in the last section. Let $\alpha > 0$:

---
[12] Hence, the reader may skip this rather technical section.

$$d_1^\alpha M := \Box_S^{\leq\alpha} M \cap \Box_H^{\leq\alpha} M$$
$$d_2^\alpha M := \Box_S^{\leq\alpha} M \cap \Box_H^{\leq\alpha} \Box_S^{\leq\alpha} M$$
$$d_3^\alpha M := \Box_S^{\leq\alpha} \Box_H^{\leq\alpha} M \cap \Box_H^{\leq\alpha} M$$
$$d_4^\alpha M := \Box_S^{\leq\alpha} \Box_H^{\leq\alpha} M \cap \Box_H^{\leq\alpha} \Box_S^{\leq\alpha} M$$

We define $\mathcal{H}_0(\mathcal{T}) := \mathcal{T}$, $\mathcal{H}_{<\alpha}(\mathcal{T}) := \bigcup_{\beta<\alpha} \mathcal{H}_\beta(\mathcal{T})$. For $\alpha > 0$ we set

$$\mathcal{H}_\alpha(\mathcal{T}) := \mathcal{H}_{<\alpha}(\mathcal{T}) \cup \bigcup_{i=1}^4 d_i^\alpha \mathcal{H}_{<\alpha}(\mathcal{T}).$$

Let $\mathcal{H}_\infty(\mathcal{T}) := \bigcup_\alpha \mathcal{H}_\alpha(\mathcal{T})$. For this construction we used the combination $\Box_X^{\leq\alpha} \Box_Y^{\leq\alpha} M$ instead of $\Box_X^{\leq\alpha}(M \cup \Box_Y^{\leq\alpha} M)$. This is justified by the following lemma which implies that $\mathcal{H}_{<\alpha}(\mathcal{T}) \subseteq \Box_Y^{\leq\alpha} \mathcal{H}_{<\alpha}(\mathcal{T})$.

Lemma A.3 $\mathcal{H}_\alpha(\mathcal{T})$ *is transitive for all α.* □

Before we prove the lemma, we note the following fact:

Fact A.4 *Let $X, Y \in \mathrm{DP}$, $X \neq Y$. Assume that $w \in \mathcal{H}_\alpha(\mathcal{T}) \setminus \mathcal{T}$, $v \in w(X)$, and $\mathrm{otp}(w) = \alpha$. Then $\mathrm{otp}(v) = \alpha \Rightarrow w \in \Box_X^{\leq\alpha} \Box_Y^{\leq\alpha} \mathcal{H}_{<\alpha}(\mathcal{T})$.*

Proof: If $w \notin \Box_X^{\leq\alpha} \Box_Y^{\leq\alpha} \mathcal{H}_{<\alpha}(\mathcal{T})$, then $w \in \Box_X^{\leq\alpha} \mathcal{H}_{<\alpha}(\mathcal{T})$. But then $\mathrm{otp}(v)$ would have to be smaller than α. ⊞

Proof of Lemma A.3: We have to show that for all α $w \in \mathcal{H}_\alpha(\mathcal{T})$ implies $w(S) \cup w(H) \subseteq \mathcal{H}_\alpha(\mathcal{T})$. By induction over α: For $\mathcal{H}_0(\mathcal{T}) = \mathcal{T}$ the claim holds by definition. Assume that $w \in \mathcal{H}_\alpha(\mathcal{T}) \setminus \mathcal{H}_{<\alpha}(\mathcal{T})$. Then, suppose $w(S) \not\subseteq \mathcal{H}_\alpha(\mathcal{T})$. Let $v \in w(S) \setminus \mathcal{H}_\alpha(\mathcal{T})$. By introspection $v(S) = w(S)$. As $w(S) \setminus \mathcal{H}_\alpha(\mathcal{T}) \neq \emptyset$, it follows that $w \in \Box_S^{\leq\alpha} \Box_H^{\leq\alpha} \mathcal{H}_{<\alpha}(\mathcal{T})$. Therefore $v \in \Box_S^{\leq\alpha} \Box_H^{\leq\alpha} \mathcal{H}_{<\alpha}(\mathcal{T}) \cap \Box_H^{\leq\alpha} \mathcal{H}_{<\alpha}(\mathcal{T}) = d_3^\alpha \mathcal{H}_{<\alpha}(\mathcal{T})$. Therefore, $v \in \mathcal{H}_\alpha(\mathcal{T})$, which contradicts the assumption. The case for $w(H)$ is symmetric. ⊞

For the proofs of the following lemmas we need one more technical fact:

Fact A.5 *Let $X, Y \in \mathrm{DP}$, $X \neq Y$. Assume that $w \in \mathcal{H}_\alpha(\mathcal{T}) \setminus \mathcal{T}$, $v \in w(X)$, and $\mathrm{otp}(w) = \alpha$. Then $\forall u \in T(v) (\mathrm{otp}(u) = \alpha \Rightarrow u \in w(X))$.* □

Proof: Let $\mathrm{otp}(u) = \alpha$, then (1) implies $w \in \Box_X^{\leq\alpha} \Box_Y^{\leq\alpha} \mathcal{H}_{<\alpha}(\mathcal{T})$. Therefore, $\forall u \in v(X) = w(X) \, \forall u' \in u(Y) \, \mathrm{otp}(u') < \alpha$. Now it follows by Fact A.2 that $u \in w(X)$. ⊞

The next lemma shows that our construction provides in step α really all possibilities with complexity α:

Lemma A.6 $\forall w \in \mathcal{H}_\infty(\mathcal{T}) \, \mathrm{otp}(w) = \alpha \Rightarrow w \in \mathcal{H}_\alpha(\mathcal{T})$. □

Proof: Assume that we have shown by induction that $w \in \mathcal{H}_{<\beta}(T)$ & $\mathrm{otp}(w) = 0$ implies $w \in T$. Suppose $w \in \mathcal{H}_\beta(T) \setminus \mathcal{H}_{<\beta}(T)$, $\mathrm{otp}(w) = 0$. Hence, $w \in \square_S^{<\beta}\mathcal{H}_{<\beta}(T)$ or $w \in \square_S^{<\beta}\square_H^{<\beta}\mathcal{H}_{<\beta}(T)$. Therefore, either $\exists v \in w(S)\, v \in \mathcal{H}_{<\beta}(T)$, or $\exists v' \in w(S)\, \exists u \in v'(H)\, u \in \mathcal{H}_{<\beta}(T)$. Due to the transitivity of $\mathcal{H}_{<\beta}(T)$ it follows that $T(v), T(u) \subseteq \mathcal{H}_{<\beta}(T)$, and $\mathrm{otp}(w) = 0$ implies that $w \in T(w) = T(v) = T(u) \subseteq \mathcal{H}_{<\beta}(T)$. Hence, it follows by I.H. that $w \in T$.

Assume that $\alpha > 0$, and that the proposition holds for all $\beta < \alpha$. Let $w \in \mathcal{H}_\infty(T)$, $\mathrm{otp}(w) = \alpha$. Then, for some $\gamma \geq \alpha$ $w \in \mathcal{H}_\gamma(T)$. Let $\gamma = \alpha + 1$. We first consider the case where $w \in \square_S^{\leq\alpha+1}\square_H^{\leq\alpha+1}\mathcal{H}_{<\alpha+1}(T)$, i.e., $w \in \square_S^{\leq\alpha}\square_H^{\leq\alpha}\mathcal{H}_\alpha(T)$. Suppose that $\exists v \in w(S)\exists u \in v(H)\, \mathrm{otp}(u) = \alpha$. Then $w \in T(u)$ by Fact 3.1, and by Fact A.4 it follows that $u \in \square_H^{\leq\alpha}\square_S^{\leq\alpha}\mathcal{H}_{<\alpha}(T)$ because $u \in \mathcal{H}_\alpha(T)$, $u \in u(H)$ and $\mathrm{otp}(u) = \alpha$. It follows from Fact A.5 that $w \in u(H)$. Hence, $w \in \square_S^{\leq\alpha}\mathcal{H}_{<\alpha}(T)$. But then it follows that $\mathrm{otp}(v) < \alpha$, and therefore $\mathrm{otp}(u) < \alpha$, in contradiction to the assumption. Hence, $\forall v \in w(S)\forall u \in v(H)\, \mathrm{otp}(u) < \alpha$. By I.H. it follows that $\forall v \in w(S)\ \forall u \in v(H)\, u \in \mathcal{H}_{<\alpha}(T)$. Together, this implies $w \in \square_S^{\leq\alpha}\square_H^{\leq\alpha}\mathcal{H}_{<\alpha}(T)$.

Now we consider the case $w \in \square_S^{\leq\alpha+1}\mathcal{H}_{<\alpha+1}(T)$. Then $w \in \square_S^{\leq\alpha}\mathcal{H}_\alpha(T)$. Let $v \in w(S)$, $\mathrm{otp}(v) = \alpha$. Then $v(S) = w(S)$ and $v \in \square_S^{\leq\alpha}\square_H^{\leq\alpha}\mathcal{H}_{<\alpha}(T)$. Hence $w \in \square_S^{\leq\alpha}\square_H^{\leq\alpha}\mathcal{H}_{<\alpha}(T)$.

The case for H is symmetric. Therefore $w \in \mathcal{H}_\alpha(T)$. For $\gamma > \alpha + 1$ the proposition follows easily with the I.H. ⊞

Now we can prove the central claim, that the hierarchy of the $\mathcal{H}_\alpha(T)$ is identical to $\mathcal{H}(T)$.

Lemma A.7 *For all α: $\mathcal{H}^\alpha(T) = \mathcal{H}_\alpha(T)$.* ⊟ *Moreover, this claim remains valid if we use only d_1^α to d_3^α in the above construction of $\mathcal{H}_\alpha(T)$.*

Proof: We prove the lemma by induction over α. We write $\mathcal{H}^{<\alpha}(T)$ for $\bigcup_{\beta<\alpha}\mathcal{H}^\beta(T)$.

For $\alpha = 0$ the claim follows by definition and Lemma A.1. Assume that $\alpha > 0$, and that $w \in \mathcal{H}^\alpha(T)$. If $\mathrm{otp}(w) < \alpha$, then $w \in \mathcal{H}^\beta(T)$ for some $\beta < \alpha$, and it follows by I.H. that $w \in \mathcal{H}_\beta(T)$. Assume $\mathrm{otp}(w) = \alpha$, hence, $w \notin T$. Lemma A.1 shows that $w \in \square_S^{\leq\alpha}\mathcal{H}^{<\alpha}(T)$, or that $w \in \square_H^{\leq\alpha}\mathcal{H}^{<\alpha}(T)$. If $w \notin w(S) \cup w(H)$, then w is an element of the intersection of both classes, and therefore we find by I.H. $w \in \square_S^{\leq\alpha}\mathcal{H}_{<\alpha}(T) \cap \square_H^{\leq\alpha}\mathcal{H}_{<\alpha}(T) = \mathrm{d}_1\mathcal{H}_{<\alpha}(T)$. Assume that $w \in w(S)$. By introspection it holds that for all $v \in w(S)$: $v \in v(S)$, hence, by Fact 3.1 and Lemma A.1 that for all $v \in w(S) \setminus T$: $v \in \square_S^{\leq\alpha}\mathcal{H}^{<\alpha}(T)$. Hence, w must be an element of $\square_S^{\leq\alpha}\square_H^{\leq\alpha}\mathcal{H}^{<\alpha}(T)$, therefore $w \in \square_S^{\leq\alpha}\square_H^{\leq\alpha}\mathcal{H}^{<\alpha}(T) \cap \square_H^{\leq\alpha}\mathcal{H}^{<\alpha}(T)$ and with I.H. it follows that $w \in \mathrm{d}_3^\alpha\mathcal{H}_{<\alpha}(T) \subseteq \mathcal{H}_\alpha(T)$. This proves that $\mathcal{H}^\alpha(T) \subseteq \mathcal{H}_\alpha(T)$, and, moreover, that d_4^α was not necessary for the definition of $\mathcal{H}_\alpha(T)$.

Next assume that $w \in \mathcal{H}_\alpha(T)$. If $\mathrm{otp}(w) < \alpha$, the claim follows by I.H. and Lemma A.6. Hence, assume that $\mathrm{otp}(w) = \alpha$, which implies $w \notin T$. We have to show that $(\mathbf{H_1})$ to $(\mathbf{H_3})$ hold.

Suppose $w \in w(S) \cap w(H)$. From $w \in w(S)$ and $\mathrm{otp}(w) = \alpha$ it follows by Fact A.4 that w must be an element of $\square_S^{\leq \alpha} \square_H^{\leq \alpha} \mathcal{H}_{<\alpha}(\mathcal{T})$, and therefore in $\mathrm{d}_3^\alpha \mathcal{H}_{<\alpha}(\mathcal{T})$. But then, $\forall v \in w(H)\, \mathrm{otp}(v) < \alpha$, which contradicts $w \in w(H)$.

Suppose next that there is a $v \in T(w)$ such that there exists a $u \in v(S)$ with $w \in T(u)$. It follows by Fact 3.1 that $\mathrm{otp}(v) = \mathrm{otp}(u) = \mathrm{otp}(w) = \alpha$. But then, it follows by Fact A.5 that $w \in v(S)$.

Finally, $w \in \mathcal{H}(\mathcal{T}) \Rightarrow w(S), w(H) \subseteq \mathcal{H}(\mathcal{T})$ follows by transitivity of $\mathcal{H}(\mathcal{T})$, Lemma A.3.

As $\mathcal{H}(\mathcal{T})$ is the largest subclass of $\mathcal{F}$ where $(\mathbf{H_1})$ to $(\mathbf{H_3})$ hold, it follows that $\mathcal{H}_\alpha(\mathcal{T}) \subseteq \mathcal{H}^\alpha(\mathcal{T})$. ⊞

Index

abstract object, 74
abstraction, 28
absurdity, *see* falsum
AC, *see* axiom of choice
accesibility relation, 146
Ackermann's star device, 35, 39
Ackermann, Wilhelm, 5, 23, 27–33, 35, 39–43, 45, 46, 48–51, 53
action
  justified, 159
  mutually justified, 160
Aczel, Peter, 6, 112, 132, 134, 136, 137, 143, 148
AFA, *see* axiom, antifoundation
AFA set theory, 2, 5, 134–137, 144–150, 167, 168, 171
algebraic specification, 112
α-congruence, 127
analysis, 12, 23, 31
analytic, 64
antinomies, *see* paradoxes
applicative structure, 127
approximation
  of a map, 97
Aristotle, 6, 24
arithmetic, 13, 23, 60, 71

consistency of, 17
  Peano, 65, 101
assertion, 143–146
  basic, 151
  felicious, 171
  mutually justified, 164
  update effect of a, 171
automated theorem proving, 59, 68, 69
axiom
  Ackermann's, 104
  antifoundation, 112, 134, 137, 143, 140, 150
  infinity, 73
  nominalization, 57
  of choice, 18, 61, 65, 79, 87, 101, 102
  of comprehension, 4, 6, 57, 60
  of determinateness, 4
  of extensionality, 4, 6, 57, 61, 66, 111, 112, 117, 118
  of foundation, 2, 5, 111, 112, 117, 137, 148
  of infinity, 57, 61, 72, 80, 87
  of replacement, 102
  of restriction, *see* axiom of wellfoundedness

  of separation, 61
  of wellfoundedness, 106
axiomatic method, 1, 12, 13, 59, 60, 62

Babbage, Charles, 67
Barwise, Jon, 60
Behmann, Heinrich, 24, 27, 31, 48
Bernays, Paul, 28, 65
Bernstein, Alex, 68
β-congruence, 127
β-conversion, 28, 127
β-rule, 127
bisimulation, 112, 135, 136
Bolzano, Bernhard, 1
Boole, George, 1, 62
Boolean (in map theory), 104
Boolos, George, 6
branch, 89–91
Brouwer, Luitzen Egbertus Jan, 3, 4, 19

Cajetan (*i.e.*, Thomas de Vio), 6
Cantor, Georg, 1, 11, 12, 14–16, 25, 26, 63, 148
  on the paradoxes, 15
Carnap, Rudolf, 73
category theory, 3

Index

chain
  infinite descending, 115
choice operator, 70, 71, 101–103
Church, Alonzo, 27, 63
Chwistek, Leon, 7
circularity, 2, 143, 147, 155, 163
class, 108
  as extension of a concept, 24–25
  dual (in map theory), 109
  logical conception, 25
  representation in map theory, 108
  vs. set (in NBG), 67
class theory, 25, 53–54
Cohen, Paul J., 14, 61
collapsing theorem (Mostowski), 135
collection
  absolute infinite, 15, 16
  inconsistent, 15
  of all cardinals, 14, 15
  of all ordinals, 15
common knowledge, 2, 144, 150, 151, 154, 156, 170, 171
compact, 97
completeness
  combinatory, 27
  deductive, 29
comprehension, 122
  principles of, 26
  unrestricted, 15, 23, 27, 53, 111, 148
concept
  ϵ-theoretic, 121–123
  instance of an ϵ-theoretic, 122, 123
  meaning (of an ϵ-theoretic con.), 122
conditional, 23, 30, 32
  classical, 30

inferential interpretation, 30, 33, 38, 39
  intuitionistic, 30
consistency, 13
  relative results, 136
consistency proof, 3, 4, 13, 17, 23, 29, 53, 136
constructivism, 3
continuity, 106–108
  κ-continuity, 107
  κ-Scott continuity, 95, 107
  ω-continuity, 107
  ω-Scott continuity, 95
  Scott continuity, 95, 107
  uniform, 107, 108
continuum hypothesis, 11, 12, 14, 18
contraction, 53
contractum, 127
contraposition, 31
cumulative conception of sets, see iterative conception of sets
cumulative hierarchy, 6, 148
cut, 41, 43, 44
cut rule, 35

Davis, Martin, 68, 69
de Bruijn, Nicolas Govert, 69
De Morgan, Augustus, 62
Dedekind, Richard, 1
deduction theorem, 29, 35, 45, 53
diagonal argument, 26
dialogue situation, 147
  basic, 144, 147
  complexity of a, 157, 158, 160
  defective, 143, 171
  ideal, 143, 147, 151, 156
  non-ideal, 151
  outer, 147, 149, 168

dilution, 43
disjunction
  parallel, 89–91
distribution (in propositional logic), 41, 44
domain, 112
  κ-Scott, 107
drawing (of a map), 96

entailment, 33, 35
epistemic perspective, 143, 151–154, 159, 166, 171
epistemic possibility, 150
epistemic update, 144
ϵ-quotient, 130
  extensional, 131
ϵ-structure, 111–113, 115, 137
  extensional, 118
  generated substructure, 119
  minimal pointed to (an element), 121
  substructure of a, 118
$\in_T$-structure, 139
epsilon operator, see choice operator
equality, 60
equivalence (ϵ-structures)
  extension-conform, 130, 136
equivalence (formulas), 32
equivalence (terms), 93
  extensional, 80
  intensional, 80
  root equivalence, 92
η-rule, 127
excluded middle, 102
existence
  pure, 89, 90, 92
exportation principle, 45
  modified, 42, 45
extension, 6, 111
  ϵ-extension, 118
  course-of-values, 25

extensionality, 53, 80, 96, 112, 130–134
   rule of (in MT), 96
   strong, 137

factorization theorem, 131
falsity conditions, 48
falsum, 28, 36, 47, 48
(FEA) (Gentzen rule), 38
Feferman, Solomon, 3, 5
(FES) (Gentzen rule), 35, 36, 38, 39, 46
finitism, 19
Fitch, Frederic Brenton, 28, 49
fixed point operator, 95
formula
   Δ_0-, 116
   ε-, 115
   absolute, 119
   homogeneous, 31, 36, 38, 40
   inhomogeneous, 31
Fraenkel, Abraham Adolf, 1, 4, 61, 62, 73
Frege, Gottlob, 1, 3, 6–8, 14, 16, 17, 25, 53, 58, 59, 62, 63, 65, 80, 148
function, 2, 3, 112
   ε-function, 124
   computable, 3, 87
   continuous, 106
   in MT, 89, 90
   partial, 90
   recursive, 71
   self-applicable, 112
   total, 90
   uniformly κ-continuous, 108
functional application, 89, 91, 127

Geach, Peter Thomas, 7
Gentzen, Gerhard, 41
Gentzen, Gerhard, 33, 35, 38
geometry, 13, 60
Gilmore, Paul, 5, 7, 57, 58, 73, 80
Gödel, Kurt, 2, 14, 26, 28, 59, 62, 65, 80
graph, 134
   accessible pointed, 134
   decoration of a, 134
   well-founded, 134
graphs
   bisimular, 135
Grishin, Vycheslav Nikolaevich, 53

Hales, Thomas, 59
Halldén, Sören, 28
Halmos, Paul, 60
Henkin, Leon, 66
Hermes, Hans, 4
Hilbert, David, 2, 11–18, 62, 63
Hintikka, Jaakko, 143
homogenity principle, 30–31, 45, 46, 50, 54
homomorphism
   ε-homomorphism, 119
   strong, 120
Husserl, Edmund, 17
hyperset theory, 115

ideal calculus, see also set theory, naive, 4–6
impredicativity, 138
inconsistency
   Post, 29
independence proofs, 118
infinite, 12
infinity, 102
information state, 147, 149
intensionality, 2, 5, 73, 80, 111, 127
introspection, 164, 167
introspectivity, 150, 163
intuitionism, 3, 12, 19

ITT, 57, 58, 73, 80

König, Julius, 18
Kant, Immanuel, 13, 64
Kepler Conjecture, 59
Ketonen rules, 36
Kripke structure, 143

ℓ, 60, 62–63
λ-calculus, 2, 28, 66, 75, 89, 92, 107, 111, 112
   ε-model of the, 126, 128, 130
   extensional model of the, 128
   model of the, 121, 128
Leśniewski, Stanisław, 7, 25
least upper bound
   of two maps, 98, 99
Leibniz, Gottfried Wilhelm, 62, 67, 77
Lewis, David, 7
logic
   $\in_T$-logic, 111–113, 137–140
   $\in_\mu$-logic, 140
   ε-logic, 112, 115–117
   classical, 45, 46, 48, 52
   combinatory, 27
   elementary, see ℓ
   exportative, 45
   first order, see also ℓ, 62, 65, 69, 72
   higher order, 57, 61, 62, 72
      completeness of, 66
   infinitary, 19
   intuitionistic, 45, 46, 48
   minimal, 31
   nominalistic, see NL
   of nonsense, 28
   partial, 28, 48
   prime-, see P
   relevance, 41–43, 45
   sequent, 23, 73, 77
   strongly exportative, 46

180   Index

supra-, *see* Q
typefree, 29
weakly exportative, 53
logicism, 6, 73
loop, 115, 125
Löwenheim, Leopold, 62

M-computer, 92
map, 88, 93
  coherent, 98
  prime, 98, 99
  wellfounded, 88, 104, 106, 108
map theory, 2, 3, 5, 24, 87–89, 94, 101–105, 109
  standard model of, 95
Markov, Andrei Andreyevich, 49
Martin-Löf, Per, 61
mathematics
  and language, 73
  axiomatization of, 58, 62–64
  formalization of, 58, 65–68
  integration of, 58, 72–80
  intensional, 5
  mechanization of, 58, 69
membership, 3, 25, 28, 53, 112
metamathematics, 12, 19, 29, 33
modelling, 2, 3, 112, 143
  with ϵ-structures, 121–130
modus ponens, 54
monotony, 103
Mostowski, Andrzej, 66
MT, *see* map theory
mutual learning, 150, 151, 171
Myhill, John, 2, 5, 26

NBG, 1, 66, 108
negation, 31, 47–50
  law of double, 32
  of compound formulas, 32
  of conditionals, 32
  strong, 28, 49
  weak, 53
Nelson, David, 49
NL, 57, 58, 63, 65, 69, 73–80
nominalism, 7, 74, 80
nominalization, 73, 76
normal form, 28, 89, 90, 92
  of NL-terms, 76
Novak, Ilse L., 66
number
  natural, 71
numbers
  cardinal, 14
  transfinite, 12, 14, 16

object variable, 89, 90

P, 63, 65, 69, 80
pair, 3, 112, 114, 123
  ϵ-pair, 124
  Kuratowski, 122, 124
paraconsistency, 69, 81
paradox
  Banach-Tarski, 102
  Cantor, 15
  Curry, 23, 24, 29, 53
  Hilbert, 16
  of the liar, 138
  Zermelo-Russell, 8, 12, 16, 17, 23, 27, 28, 48, 53, 58–61, 63, 80, 117
paradoxes
  of implication, 45
  of set theory, 1, 11, 12, 15, 23, 26
  as use mention confusions, 73
partiality, *see* truth value gap
Peano, Guiseppe, 63
Peirce's law, 46
perspectival operator, 159, 166
platonism, 7
P/NP-problem, 72
Porphyry, 6
possibilities, 146–151, 154, 168
possible derived extension (of a class of possibilities), 166
possible world, 143, 146, 147
predicates
  partial, 28
predication, 25
predicativism, 3
Presburger, Mojżesz, 69
programming, 88, 112
  functional, 127
proof theory, 12
proposition, 64, 138, 139
propositions-as-types, 61

Q, 63, 65, 69–72, 80
QED Manifesto, 72
QND, *see* Quartum Non Datur
quantification, 28, 50–52, 88, 101–104
  axioms in map theory, 104
  bounded, 119
  domain of, 88, 101, 103, 104
  impredicative, 111
  over propositions, 112
quantifier, 101
  existential, 50, 101
  lazy, 103
  strict, 103
  universal, 30, 50, 101
  over types, 70
Quartum Non Datur, 94
Quine, Willard Van Orman, 64, 65

rationality constraint, 155, 163, 166
reasoning
  as computation, 67

extensionality, 53, 80, 96, 112, 130–134
  rule of (in MT), 96
  strong, 137

factorization theorem, 131
falsity conditions, 48
falsum, 28, 36, 47, 48
(FEA) (Gentzen rule), 38
Feferman, Solomon, 3, 5
(FES) (Gentzen rule), 35, 36, 38, 39, 46
finitism, 19
Fitch, Frederic Brenton, 28, 49
fixed point operator, 95
formula
  Δ_0-, 116
  ε-, 115
  absolute, 119
  homogeneous, 31, 36, 38, 40
  inhomogeneous, 31
Fraenkel, Abraham Adolf, 1, 4, 61, 62, 73
Frege, Gottlob, 1, 3, 6–8, 14, 16, 17, 25, 53, 58, 59, 62, 63, 65, 80, 148
function, 2, 3, 112
  ε-function, 124
  computable, 3, 87
  continuous, 106
  in MT, 89, 90
  partial, 90
  recursive, 71
  self-applicable, 112
  total, 90
  uniformly κ-continuous, 108
functional application, 89, 91, 127

Geach, Peter Thomas, 7
Gentzen, Gerhard, 41
Gentzen, Gerhard, 33, 35, 38
geometry, 13, 60
Gilmore, Paul, 5, 7, 57, 58, 73, 80
Gödel, Kurt, 2, 14, 26, 28, 59, 62, 65, 80
graph, 134
  accessible pointed, 134
  decoration of a, 134
  well-founded, 134
graphs
  bisimular, 135
Grishin, Vycheslav Nikolaevich, 53

Hales, Thomas, 59
Halldén, Sören, 28
Halmos, Paul, 60
Henkin, Leon, 66
Hermes, Hans, 4
Hilbert, David, 2, 11–18, 62, 63
Hintikka, Jaakko, 143
homogenity principle, 30–31, 45, 46, 50, 54
homomorphism
  ε-homomorphism, 119
  strong, 120
Husserl, Edmund, 17
hyperset theory, 115

ideal calculus, see also set theory, naive, 4–6
impredicativity, 138
inconsistency
  Post, 29
independence proofs, 118
infinite, 12
infinity, 102
information state, 147, 149
intensionality, 2, 5, 73, 80, 111, 127
introspection, 164, 167
introspectivity, 150, 163
intuitionism, 3, 12, 19

ITT, 57, 58, 73, 80

König, Julius, 18
Kant, Immanuel, 13, 64
Kepler Conjecture, 59
Ketonen rules, 36
Kripke structure, 143

ℓ, 60, 62–63
λ-calculus, 2, 28, 66, 75, 89, 92, 107, 111, 112
  ε-model of the, 126, 128, 130
  extensional model of the, 128
  model of the, 121, 128
Leśniewski, Stanisław, 7, 25
least upper bound
  of two maps, 98, 99
Leibniz, Gottfried Wilhelm, 62, 67, 77
Lewis, David, 7
logic
  $\in_T$-logic, 111–113, 137–140
  $\in_\mu$-logic, 140
  ε-logic, 112, 115–117
  classical, 45, 46, 48, 52
  combinatory, 27
  elementary, see ℓ
  exportative, 45
  first order, see also ℓ, 62, 65, 69, 72
  higher order, 57, 61, 62, 72
    completeness of, 66
  infinitary, 19
  intuitionistic, 45, 46, 48
  minimal, 31
  nominalistic, see NL
  of nonsense, 28
  partial, 28, 48
  prime-, see P
  relevance, 41–43, 45
  sequent, 23, 73, 77
  strongly exportative, 46

180   Index

supra-, *see* Q
typefree, 29
weakly exportative, 53
logicism, 6, 73
loop, 115, 125
Löwenheim, Leopold, 62

M-computer, 92
map, 88, 93
  coherent, 98
  prime, 98, 99
  wellfounded, 88, 104, 106, 108
map theory, 2, 3, 5, 24, 87–89, 94, 101–105, 109
  standard model of, 95
Markov, Andrei Andreyevich, 49
Martin-Löf, Per, 61
mathematics
  and language, 73
  axiomatization of, 58, 62–64
  formalization of, 58, 65–68
  integration of, 58, 72–80
  intensional, 5
  mechanization of, 58, 69
membership, 3, 25, 28, 53, 112
metamathematics, 12, 19, 29, 33
modelling, 2, 3, 112, 143
  with ε-structures, 121–130
modus ponens, 54
monotony, 103
Mostowski, Andrzej, 66
MT, *see* map theory
mutual learning, 150, 151, 171
Myhill, John, 2, 5, 26

NBG, 1, 66, 108
negation, 31, 47–50
  law of double, 32

of compound formulas, 32
of conditionals, 32
  strong, 28, 49
  weak, 53
Nelson, David, 49
NL, 57, 58, 63, 65, 69, 73–80
nominalism, 7, 74, 80
nominalization, 73, 76
normal form, 28, 89, 90, 92
  of NL-terms, 76
Novak, Ilse L., 66
number
  natural, 71
numbers
  cardinal, 14
  transfinite, 12, 14, 16

object variable, 89, 90

P, 63, 65, 69, 80
pair, 3, 112, 114, 123
  ε-pair, 124
  Kuratowski, 122, 124
paraconsistency, 69, 81
paradox
  Banach-Tarski, 102
  Cantor, 15
  Curry, 23, 24, 29, 53
  Hilbert, 16
  of the liar, 138
  Zermelo-Russell, 8, 12, 16, 17, 23, 27, 28, 48, 53, 58–61, 63, 80, 117
paradoxes
  of implication, 45
  of set theory, 1, 11, 12, 15, 23, 26
  as use mention confusions, 73
partiality, *see* truth value gap
Peano, Guiseppe, 63
Peirce's law, 46
perspectival operator, 159, 166

platonism, 7
P/NP-problem, 72
Porphyry, 6
possibilities, 146–151, 154, 168
possible derived extension (of a class of possibilities), 166
possible world, 143, 146, 147
predicates
  partial, 28
predication, 25
predicativism, 3
Presburger, Mojżesz, 69
programming, 88, 112
  functional, 127
proof theory, 12
proposition, 64, 138, 139
  propositions-as-types, 61

Q, 63, 65, 69–72, 80
QED Manifesto, 72
QND, *see* Quartum Non Datur
quantification, 28, 50–52, 88, 101–104
  axioms in map theory, 104
  bounded, 119
  domain of, 88, 101, 103, 104
  impredicative, 111
  over propositions, 112
quantifier, 101
  existential, 50, 101
  lazy, 103
  strict, 103
  universal, 30, 50, 101
  over types, 70
Quartum Non Datur, 94
Quine, Willard Van Orman, 64, 65

rationality constraint, 155, 163, 166
reasoning
  as computation, 67

redex, 127
reduction, 90
  rule of, 90
reduction sequence, 90
reduction step, 90
relation, 112
  ϵ-relation, 124
  non-well-founded, 126
  partially defined, 27
  well founded, 120
replacement of equivalents, 32
residuation, 45
Rosser, John Barkley, 61
rule of inference, 33
  homogeneous, 36
Russell, Bertrand, 1, 11, 17, 18, 27, 57, 58, 61–65, 73

Schmidt, Arnold, 30, 38
Scholz, Heinrich, 4
Schröder, Ernst, 1, 6
Scott model, 88, 112
Scott order, 95
self application, 2, 127
self-referentiality, 2, 111, 143
semantic level, 30, 36, 48
semi-intuitionism, 3
set
  ϵ-set, 2, 111–113, 121, 125, 126, 136, 137
  Aczel set, 112
  as collection, 3, 6, 15, 25, 148
  as extension of a concept, 6, 25, 148
  as structured object, 5, 148
  finished, 14
  finite, 87
  functional-reflexive ϵ-set, 126
  intensional, 5, 112
  internal structure, 149
  iterative conception, 6, 24, 25, 27, 61, 148
  logical conception, 6, 24–25
  non-wellfounded, 7, 111, 112, 123, 134, 143, 146, 148, 149
  picture of a, 134
  transfinite, 15
  unintended, 11, 12, 14–17, 26
set formation, *see* comprehension
set theory, 72
  Ackermann's, 24
  axiomatization of, 1, 16–18
  naive, 12, 58, 60, 73
  undecidable small fragments of, 66
set-structure, 113, 114
  transitive, 113, 122, 125
sets
  and extensions in ZFC, 118
  criterion of identity for, 4, 5
  representation of (in map theory), 105
Shannon number, 68
Shannon, Claude, 66, 68
Shoenfield, Joseph R., 61, 62
Σ-logic, 33–47
  homogeneous, 33–42
Σ-sequent, 35
sincerity condition, 146, 163, 170
singleton, 122
Skolem, Thoralf, 1, 19, 61
Skolemism, 19
small
  V-small, 107
solution lemma, 148, 168
speaker's intention, 143, 147, 154–155, 159, 162, 171
speech act, 171
Strachey, Christopher, 68
synthetic, 64
system of equations, 148, 150, 167
Szmielew, Wanda, 66

Tarski biconditional, 112
Tarski, Alfred, 66
term, 89
  bottom, 92
  function, 92
  reducable, 90
  true, 92
Tertium Non Datur, 27
transitivity
  of a class, 150, 163
truth predicate, 112, 138
  total, 111, 138
truth value gap, 23, 31, 32
Turing machine
  universal, 67
Turing, Alan, 66, 67
type
  transfinite, 69, 70, 73, 80
type theory, 23, 27, 61–63, 80
  vs. set theory, 57, 59–62, 66, 81
  constructive, 61
  intensional, *see* ITT
universe
  antifounded, 135
update, 143, 153
  fixed-point of a, 165
  normal mutual, 150, 151, 163, 164, 168
  successful, 153
urelement, 89, 90, 148, 167

verum, 31, 36
verum ex quodlibet, 45
von Neumann, John, 1, 3, 62, 65, 67

Wang, Hao, 61, 65, 66
well-ordering theorem, 14, 18

Weyl, Hermann, 3, 6, 7, 60
Whitehead, Alfred North, 18, 62
Wiener, Norbert, 66
William of Ockham, 74

Zermelo, Ernst, 1, 2, 6, 11, 12, 16–18, 27, 61–63, 65, 102
    axiomatization of set theory, 12
    on Skolemism, 19
    on the paradoxes, 19
ZF, 24, 61
ZFC, 1, 5, 61, 87, 88, 93, 101, 102, 105, 106, 108, 110, 113, 117, 125, 127, 135, 137
    consistency of, 136
    model of, 135
ZFC$^-$, 135, 136
ZFC$^-$+AFA, 134, 136, 137
ZFC+SI, 88
Zorn's lemma, 102
Zuse, Konrad, 67